Design, Construction and Uses of Trawl Fishing Gear

THE AUTHOR

Dr. K.P. Biswas, M.Sc., Ph.D., D.F.Sc. (Bombay), E.F. (West Germany), F.Z.S., F.A.B.S. (Kolkata), Former Joint Director Fisheries (LI), Government of Orissa, Director Fisheries, A&N Islands, Government of India, Principal, Fisheries Training Institute, Fishery Technologist (ICAR), and at present Faculty Member of Marine Science Department, University of Calcutta and West Bengal University of Animal and Fishery Sciences.

The author operated 28 wooden fishing trawlers (32 - 48 feet LOA) from Paradeep Orissa during 1968-69 and four 57 feet steel polish trawlers during 1971-72 in Odisha coast of Bay of Bengal and is conversant with trawl fishery in Indian coast.

Design, Construction and Uses of Trawl Fishing Gear

–Author–

K.P. Biswas

2015

Daya Publishing House®

A Division of

Astral International (P) Ltd

New Delhi 110 002

Cataloging in Publication Data—DK
  Courtesy: D.K. Agencies (P) Ltd. <docinfo@dkagencies.com>
Biswas, K. P. (Kamakhya Pada), 1936- **author.**
Design, construction and uses of trawl fishing gear / author, K.P. Biswas.
  pages   cm
Includes bibliographical references and index.
ISBN 978-93-5130-670-2 (International Edition)

  1. Fisheries--India--Equipment and supplies--Design and construction.
        2. Trawls and trawling--India--Design and construction. 3.
        Fishery technology--India.   I. Title.

DDC 639.20954   23

Published by	:	**Daya Publishing House®**
		A Division of
		Astral International Pvt. Ltd.
		– ISO 9001:2008 Certified Company –
		4760-61/23, Ansari Road, Darya Ganj
		New Delhi-110 002
		Ph. 011-43549197, 23278134
		E-mail: info@astralint.com
		Website: www.astralint.com
Laser Typesetting	:	**Classic Computer Services,**
		Delhi - 110 035
Printed at	:	**Thomson Press India Limited**

PRINTED IN INDIA

Dedicated to...

Mrs Manju Biswas

Preface

The area of Fisheries Science that has contributed to the maximum of the outstanding increase in the productivity of fishing operation is fishing gear technology. As elsewhere in the world, this is true in the development of marine fisheries in India too.

Among the fishing gear, the net occupies the prime position in catching fish commercially for mass production. The use of net as the most efficient fishing gear has to be perfected for adopting wherever necessary, to the fishing conditions usually encountered. In the changing situations it might become necessary to incorporate modifications of varying degree to the existing patterns or to the designs altogether new models. In doing so the aim should be to produce a net which will give the largest catching potential for the smallest cost in materials and the least energy output. Many different factors have to be taken into account, such as, resilience of webbing, strength and elasticity, resistance to flow of water, weight and bulk, speed of operation, cost of material, conditions of fishing ground etc.It is also fundamental to find a method of determining objectively the most suitable elements of the fishing gear taking into account the behavior of fish and technical condition of operation. Again, while using different designs, the relative efficiency and selectivity of the gear must be of prime consideration. By selectivity, it is meant how much will a certain gear catch, what species and what sizes?

By its continuing efforts and from its accumulating experiences, this communication earnestly hopes to throw brighter light through its future publications to the intricacies involved in gear technology for bigger ocean fishing vessels and factory trawlers. Attempts have been made to derive clear pictures on the relative efficiency and selectivity of specific gears.

In India Exploratory Fisheries Project has the unique distinction of inducting trawl gear as the most successful gear and bottom trawling *is* the most effective fishing method in this country since early 1950s. The combination of this system revolutionized the prawn fisheries. Exploratory Fisheries Project have operated a large range of size and designs of trawl gears for shrimp and fish over the years. Efforts are made through this publication to make available the data concerning these gears to the end users. It is likely that the presentation of various data concerning the gears is not complete. But if the information furnished become useful in some manner, it would have achieved one of the aims of this book.

K. P. Biswas

Introduction

Fishing gear is the common term for the tool or tackle employed for catching fish either from a craft or from the shore of water mass. Based on the mode of operation the gear may be active like trawl or purse-seines or may be passive like stake nets, gill nets and traps. Based on its action on the fish, the gear may be entangling (gill nets), engulfing (trawl) and hooking (angling) or stupefying (electrical gears) or wounding (spears, spikes etc). Based on the material used for its construction, the gear may be webbed or non-webbed.

The fishing industry, in course of time has developed into a target fishing sector using resource specific fishing gear in developed countries much earlier than Indian coasts. Lately that spell of development also reached the Indian coast. The trawls in Andhra Pradesh are now broadly classified into three types based on the target groups, namely, fish trawls targeted for fin fish, shrimp trawls for shell fish and cephalopod trawls for exploitation of cuttle fish and squid. Shrimp trawls have undergone several changes in the course of time with increase in number of seams from two to six, increase in vertical height and also increase in length of the net. Mesh size in fish trawls in the fore part of the net increased from 150 to 2000 mm to reduce the drag. Two seam fish trawls with head rope length ranging from 20 to 84 meters are widely used for exploitation of fin fish in Andhra Pradesh. The cod end mesh size of all the trawls were less than those stipulated in the regional Marine Fishing Regulating Act, 1955.

The exploitation of the inshore fisheries using indigenous craft and gear is being intensively carried out from very ancient times. Even though, exploratory and experimental trawling operations along the Indian coast dates back to 1900 (Chidambaram, 1952), organized and systematic attempts of trawling in inshore waters has come into vogue only within last five decades or so. The exploratory trawl surveys conducted in the country can be conveniently grouped into two distinct periods, (i) those operations carried out in the pre-war days and (ii) those of the post-war period. All these surveys have indicated the presence of good fishing grounds along the Indian coast These surveys formed the basis for the development of mechanized trawl fisheries in India.

The steady increase in the number of small and medium sized mechanized vessels in India during 1960-1970 amply shows that trawling has become an established method of fishing along the Indian coast. The accent during this period was for exploiting the shrimp resources along south west and south east coasts and naturally the fishing gear developed was mainly shrimp trawls and consequently the necessity to evolve a basis for the designs presented itself. Realizing this, the trawls were subjected to detailed studies with a view to arrive at certain empirical relationships, which could form the basis for the designs of trawls. Such attempts in establishing relationships have been made by Miyamoto (1959) also.

Two seam fish trawls dominated the fishing industry until 1950s. During the next two decades, though popularization of trawling brought about a greater use of four seam nets along the west coast, two seam nets continued its popularity along the east coast (Satyanaryana *et al,* 1072). This trend still continues. It was observed that shrimp trawls have undergone several changes in the course of time, like, increasing the number of seams, from two to six. Results indicated that the mesh size in the fore part of the net increased from 150 to 2000 mm in the fish trawls, the advantage being a reduction in the drag.

Four seam and six seam shrimp trawls gained popularity along the east coast. The horizontal opening of conventional two seam trawl was invariably higher than four seam trawl, which indirectly indicated that four seam net obtained more vertical spread, which accounted for better catch of bottom fishes. Further, no significant difference in the catch rate was reported, but a striking difference in the catch composition between two seam and four seam trawls were observed. It was evaluated that the double trawl net was effective in targeting bottom as well as off bottom fishes simultaneously. The fuel consumption rate was significantly less in rope trawl while catch was significantly more.

Two seam fish trawls are widely used for exploitation of finfish resources. Shrimp trawls have undergone several changes in course of time, namely, increasing in number of seams from two to six, increase in vertical height and length of net. In order to manage the trawl resources efficiently, caution needs to be exercised in the use of trawl nets. Selective fishing practices and mesh size regulation need to be strictly adhered to for longer sustainability of fishery resources.

A technology is said to be successful, only when majority of the end users adopt it without any inhibition and gets satisfied with the result. So by studying the innovation decision process, the percolation of technologies passing through knowledge persuation, decision, implementation and confirmation stages could be tracked. If the stage(s) where the innovation decision cease is identified, appropriate follow-up measures could be taken up to make the innovation reach the confirmation stage. The findings on characteristics of users (trawl nets for fishing) influencing the innovation decisions (improved trawl design using BRDs to exclude juveniles and non-target species) would help in designing a suitable extension strategy, according to their innovation decision behavior, socio-personal profile and resource base. The findings on constraints might help to formulate appropriate remedial measures for improving the technology development, technology transfer and innovation decision processes in trawl fisheries.

The innovation decision process is efficient pertaining to the technologies which are directly related to increasing production, labor efficiency, fuel efficiency, reducing the operational expenditure and increasing the income. In the case of technologies pertaining to the conservation of resources in the interest of sustainability and environmental impact, the innovation decision efficiency scores are relatively low.

The new approach in trawling gear research includes as an important part the study of the fishing gear. Electronic telemetering instruments have proved to be the primary requisites for the study of the hydrodynamic parameters of the trawling gear. For an effective study and the development of more efficient trawling gear, the field of instrumentation has to be further extended with a view to feeding the information to the modern high speed computers for speedy and accurate analysis of the data. Just tike other fields of physical sciences, there are all possibilities of studying the basic properties and parameters of the trawling gear and developing the same with the help of electronic instrumentation, controls and data processing.

K. P. Biswas

Contents

1. **Evolution of Trawl Fishing Gear**   **1**

   Parts *of trawl net, Webbings, Square, Bosom, Jibs, Quarters, Side panels. Bellies, Throat, Cod end, Flapper, Apron, Lines and ropes. Foot rope, Bolch line, Details of construction-Baiting and creasing. Point and Bar cutting. Mesh and bar cutting, Joining of panels and take up ratio, Joining ratio, Mounting of netting. Hanging coefficient.*

   Description of trawl gear: *Conventions adopted in data sheets. Webbing, Material, Type of knot, Preservation, Color, Twine size, Breaking strength. Twine surface area. Stretched mesh, Upper edge, Lower edge. Depth, Baiting rate. Take up. Quantity, Hanging, Lines and rope-Material, Construction, Diameter, Breaking strength, conventions adopted in drawings.*

   Data and construction details: *12 m two seam fish trawl, 14 m two seam fish trawl, 15 m two seam fish trawl, 16.5 m two seam fish trawl, 20 m two seam fish trawl, 24 m two seam fish trawl, 24 m modified two seam fish trawl, 30 m two seam fish trawl, 35 m two seam fish trawl, 45 m two seam fish trawl, 17 m four seam shrimp trawl, 18 m four seam shrimp trawl, 27 m four seam shrimp trawl, 28 m four seam shrimp trawl, 43.6 m four seam shrimp trawl.*

   Otter boards: *Flat otter board 120 kg for 28 m shrimp trawl. Oval flat otter board 50-55 kg for 42-56 HP trawlers. Oval flat otter board 150 kg for 153-200 HP trawlers, Oval flat otter board 180-190 kg for 200 HP trawlers. Oval flat otter board 250 kg for 240-300 HP trawlers, Otter trawl accessories- Floats, Sinkers, Pennants, Bridles, Trawl warps, Bobbins, Tickler chain, Knotless webbing, Preservation of nets.*

2. **General Principles of the Design of the Trawl Gear**   **51**

   Classification of trawl net. Parts of trawl Net, Trawl net design and development, Developing conventional otter trawl net Size of trawl gear matching to engine horse power, Design features of trawl operating in Gulf of Mannar-Fish trawl and Coastal shrimp trawl.

3. **Working Principles and Performances of Trawl Nets**   **82**

   Instrumentation, Telemetering instruments. Continuous multi-channel data acquisition system. Data processing with the help of computer.

4. **Trawl Net Design and Development**   **85**

   Development and trial of bulged belly and Six seam trawl net. Details of 25 m bulged belly trawl. Large mesh high opening fish trawl. Belly depth studies of shrimp trawl, Six seam otter trawl, Three panel double trawl net, Overhang for trawls.

5. Mid Water Trawl One Boat Mid Water Trawl   **97**

One boat mid-water trawl. Features, Babylon, Construction, Hanging, Rectangular mid-water trawl, Oneboat mid water trawling trials in SW coast of India, One boat trawl for pelagic stocks, 10.5 m mid-water trawl, Pair trawling-fishing gear, rigging, operation.

6. Bottom Trawl Large Meshed Bottom Trawl   **111**

Large mesh bottom trawl, Design criteria of bottom trawl for 16.5 m steel trawlers. Bottom trawl trials off Tamil Nadu, Design diagram and Bottom trawls

7. Semi-Pelagic Trawl   **121**

51 m long-wing semi-pelagic trawl.

8. Shrimp trawl   **124**

Belly depth for shrimp trawls.

9. Trawling Technology Transfer to the Users   **126**

Design aspects of four seam trawl. Innovation Decision of the end users.

10. Trawl Selectivity on the Catch   **139**

Selectivity on the species. Selectivity in respect of Silver pomfret, Square mesh cod end selectivity, Cod end selectivity of Torpedo Scad.

11. Accessories of Trawl Net   **145**

Otter doors, Otter boards of different shapes

12. Sieve trawl – By catch Reduction Device   **153**

By catch Reduction Device, Sieve net design, By catch characteristics of shrimp trawl. List of species occurring in trawl catch. By catch Reduction Device, Soft By catch Reduction Device.

13. Power Requirement For Trawl Net Operation   **181**

To operate auxiliaries in fishing Trawlers, Hydraulic system and mechanical drive.

14. Electrical Trawl   **185**

Impulse generator for electrical trawl, Shape of electrodes in Sea fishing trawl.

15. Instrumentation of Trawl Gear   **192**

Under water measurement. Measurement of working depth of the trawl gear. Warp tension measurement, Measurement of angle of attack, Tilt measurement. Vertical opening, Horizontal opening, Water current, Mesh size variation. Water flow inside and outside trawl net. Other parameters.

16. Terminology for Trawl Gear Technology   **202**

Assembling the net, Assortment, Backstrop, Bar, Bar cut, Bar length, Belly, Belly line, Bolch line, Bosom, Bridle, Buoy, Buoyancy, Bytterfly, Baiting, Beam trawl. Braiding, Cod end, Cod line, Combination rope, Cork line, creasing, Danleno assembly, Denier, Depth of panel, Double mesh. Elasticity, Elongation, Eye-splice, False belly. Fiber, Fishing gear, Flapper, Figure-eight-link, Foot rope. Fly mesh, Float, G-link assembly. Gallows, Ground rope, Hanging the net, Hanging coefficient, Head line elevator. Head rope, Hemp, Jib, Jute, Kelly's eye, Kapron, Knot, Lacing, Manila, Mesh, Mesh-size, Monofilament, Multifilament, Needle, Mending, Net, Net knotless, Nylon, Otter board, Otter trawl. Ply. Point, Preservation of net, Quarter rope, Reef knot, Reel, Rotting, Seam, Seam line, Selvedge, Setting knot, Silk, Sisal, Sling, Splicing, Spreader,

Square, Strand, Sweep line, Strop, Take up ratio, Tenacity, Tensile strength, Terylene, Thread, Tickler chain. Trawl net. Twine, Twist, Trawl winch, Warp, Webbing, Wings, Vertical opening, Yarn, Yarn count.

17. Fish behavior to Approaching Trawl Gear   **207**

In approaching trawl gear, Fish reaction in the lightly rigged approaching trawl. Fish reaction to heavily rigged bobbin trawl.

18. Engineering Performance of a Trawl Gear   **212**

The flume tank, Description and Operation, Models, Technical specification of flume tank.

19. Use and operation of trawl Gear   **218**

The vessel that operate trawl gear. Trawler requirements,. Propeller nozzles. Winches, Otter doors. Trawl gallows, Warp arrangements, Small otter trawls, Structure of fishing gear. Features of net, Boat types operating the trawl net, Trawl designs in small scale mechanized fishing sector, Pomfret fish trawl, Cuttlefish trawl, Shrimp trawl, Modification in trawl.

Index   **237**

1 | Evolution of Trawl Fishing Gear

Man from the very beginning of his history had a lust for fish and developed various devices for capturing fish. The first implement probably used by man some 25000 years ago were clubs, spears, harpoons, bows and arrows. In his quest for fish, man tried to improve these extremely primitive implements and in course of time, advanced types of gears came into existence.

Among the fishing nets, one of the earliest form used was a stow net, which was fixed by two poles driven into the sea bottom and two horizontal poles tied to the net to form the mouth. This method, being entirely at the mercy of the current, was a passive one, and the escapement of fish was very much. The search for a more effective method to improve the efficiency of fish catching ultimately caused the use of drag net. Net in the form of a bag with proper mouth and wings on the either side to guide the fish, when towed along the sea bottom was found very effective to catching demersal fishes. With the availability of steam and motor power the fishing operations were extended from inshore areas to offshore areas and bigger sized nets were used. The consequent enlargement of mouth frame structure made the operation of the net more cumbersome and therefore, the rigid frame was replaced by a simpler construction, consisting of one horizontal beam, secured at each side on top of the iron shoe provided. Fishing with this beam trawl, which is often referred to as the forerunner of modern deep-sea trawl gear, was the principal method of harvesting demersal fishes in use till the turn of 19th century and still in use in North Sea and Baltic for fishing crabs, shrimps and soles.

Though introduction of beam trawl was a tremendous advancement in man's strive to devise improved types of fishing gear, the relatively slight strength of the beam and difficulty in handling precluded any suggestion of increase in its length and further enlargement of the trawl became impossible. Moreover, the vertical opening of the trawl also was very limited, confined to a few feet only. It was then necessary to abandon the rigid framing of the trawl mouth and, replace it by a non-rigid device. Attempts continued in this direction with the expansion of knowledge, experience in gear technology. Otter trawls were introduced in 1880's in which a pair of boards achived the horizontal opening and buoyant devices on head line and sinkers on ground rope maintained the vertical opening.

Having established this basic patterns, the next attempt was directed towards improving the design of the net and accessories so as to catch the maximum quantity of fish per unit effort. The introduction of Vigneron Dahl gear during 1920s was the most major step in this line. Addition of bigger sized vessels in the fishing fleet in many countries demanded nets of larger dimensions and then emerged the modified Vigneron Dahl gears and subsequently the enlarged Granton trawl patterns. These Granton trawl nets or their modified versions continued to be widely used in almost all countries in the world fisheries. The need for developing trawl net patterns different from the conventional designs suiting the local conditions was felt and different new designs were evolved in various countries of which the high opening type of Russia, the four seam pattern of Japan, the Atlantic Western trawl of Canada and the semi-bottom type of Mexico became very popular. Contribution of North West European countries in this field has been quite substantial.

Though the last two centuries witnessed phenomenal development in trawl gear practices, the initial stages have been based simply on common sense and visual observations. There did not exist an intelligent communication system between the gear technologists and fishermen due to which the long term experience of practical fishermen could not be theoretically generalized and scientifically modified. But the last seven to eight decades have been the introduction of engineering theories and application of hydro-dynamic principles coupled with systematic testing of gear to determine the factors which influence the size of catch. Studies embracing the whole complexity of hydro-dynamic behavior of trawl boards, the net, floats and other accessories were undertaken in many test houses in different parts of the world. During the last few decades, research and development of the trawl gears have assumed greater importance in India and integration of the technological approach to the actual field have given rise to diverse patterns and designs of trawl gear suited for operation in the Indian waters.

Parts of trawl net : The different parts of a trawl net can be grouped under two major heads, namely, (a) webbings and (b) lines and ropes.

Webbings Square : Square is the front portion of the upper section of a trawl which is fitted between the body and the two upper wings so that it partially overhangs the lower part of the net. The square or overhang panel extends from the upper belly to the head rope.

Wings : Wings are the forward extension of webbings on either side forming major part of the trawl mouth for guiding the fish towards the bag of the net. They are in pairs, one on either side. The upper wings are attached on either sides of square piece and lower wings on either sides of lower belly. The head rope is attached from one top wing and across the center part of the square and along the opposite top wing to the end. Rigging of front rope to the lower wing also is similar except that the square attached is replaced by the lower belly.

Bosom : Bosom is the center portion of the trawl between the wings on upper and lower sections.

Jibs: Jibs are triangular pieces of webbing attached to either sides of the upper and lower bellies at their junction with wings to present a smooth shaping to the mouth of the net. These are made in pairs, one for each side. The four seam type of trawls are invariably provided with jibs.

Quarters: Quarters are the two junctions when the top wings join the square.

Side panels : Side panels are two identical pieces of webbings, attached on either sides of the belly to join the upper and lower portion of a four seam trawl.

The portion of the webbing, that comes above the belly is termed as "top wedge". and the portion placed adjacent to belly is known as "lower wedge" or "side wedge".

Bellies : The bellies, upper and lower, form the channel of trawl through which fish moves to the cod end. Upper belly is the portion of webbing between the square and throat (or cod end) on the upper side of the trawl. It is also called "top body" or "baiting". Lower belly is the section of webbing that forms the lower body of the trawl from the foot rope to the fore part of the throat or cod end.

Throat : Throat is the portion of webbing placed in between or intermediate to the belly and cod end. It is also known as "lengthener" or "extension piece".

Cod end : Cod end is the narrow rectangular end section of the trawl net usually of heavy construction with small meshes where fish is collected during the operation of the net. One edge is joined to the end of the belly or the lengthener and the other end is laced to a thin rope, which in turn is secured to a wire rope. A cod line is heaved through for joining two sections into a bag on releasing of which the catch will be unloaded on board.

Flapper : Flapper is a small trapezoidal piece of netting whose wide front edge is laced in the fore part of the cod end to the upper panel, while the short rear edge remains free and forms the mouth of the cod end. It acts as a safety device to prevent the escape of fish from the cod end. Flapper pocket, valve, traps etc are some other names used for the flapper. In nets where lengtheners are provided, flapper is very seldom used.

Apron : Otherwise known as "Hula skirt" is an old piece of thick netting attached around the cod end as a chafing gear. It is fitted purely as a protection against wear and should not be attached so that it prevents the escape of small and immature fish.

Lines and ropes : Head rope is the rope or line forming the upper lip of the trawl to which the upper edge of the net is finally attached. Eye splices usually reinforced by thimbles and provided at both the ends of head line and floats are attached to this rope for keeping the net buoyant.

Foot rope : This rope otherwise known as ground rope is the one to which the lower edge of the net is finally attached. Splices are provided as in head rope and weights are attached to this rope for stretching the net downwards.

Bolch line: Bolch line is a line or thin rope to which the webbing is initially hung, prior to the rigging of the net to head rope and foot rope.

Belly line : Belly of trawl net is vulnerable spot and required reinforcement. Belly lines are the strengthening ropes seized along the joinings where the upper and lower panels are laced together or laced with the side panels. Belly line legs are two in number, usually provided in two seam trawls, run from the quarter mesh at foot rope along a halfer to the either end of bosom where they join the belly line. They help to bear the lateral strain on the trawl when it is being towed and hauled.

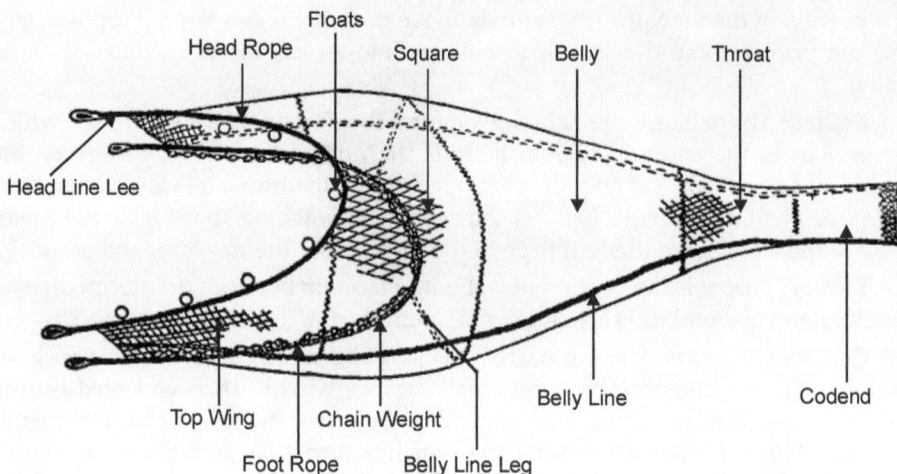

Fig. 1: Sections of a two seam trawl

Details of construction Baiting and creasing : The shape of the piece of netting of which a gear consists is achieved by reducing or increasing the number of meshes in width or length. While preparing webbings by hand braiding, the required taper is effected by increasing or decreasing the number of meshes in the concerned rows. But at present large scale manufacture of nets is undertaken, which is considerably economical than hand braiding and in such instances ready made webbings are tailored into panels of required size and shape. The cutting of webbing should be done with minimum possible wastage and for this a clear understanding of the process involved become essential. Baiting is the term generally applied for the process of cutting to obtain the desired tapering panel from the hand made or machine made webbing.

Basically there are three cutting patterns (Fig. 2a).

"All points" or the "normal" cut where the cut is perpendicular to the general course of the yarn in knotted netting. This method also known as "square cutting" or "mirror cutting" will give rectangular or square pieces of webbing,

"All bars" where the cut is parallel to a line of sequential mesh bars. Here the webbing is cut diagonally by cutting only one leg from each knot and a right angled triangular piece will be obtained where the two sides are equal. This method is sometimes referred to as "cutting by cross lines".

"All meshes" or the "transversal" cut where the cut is parallel to the general direction of the yarn in knotted netting.

The other cutting patterns are combinations of these three basic patterns, calculated according to the requirements. If the taper required is less rapid then " all bar" than "point and bar" is used, if more rapid then "mesh and bar" cutting is used.

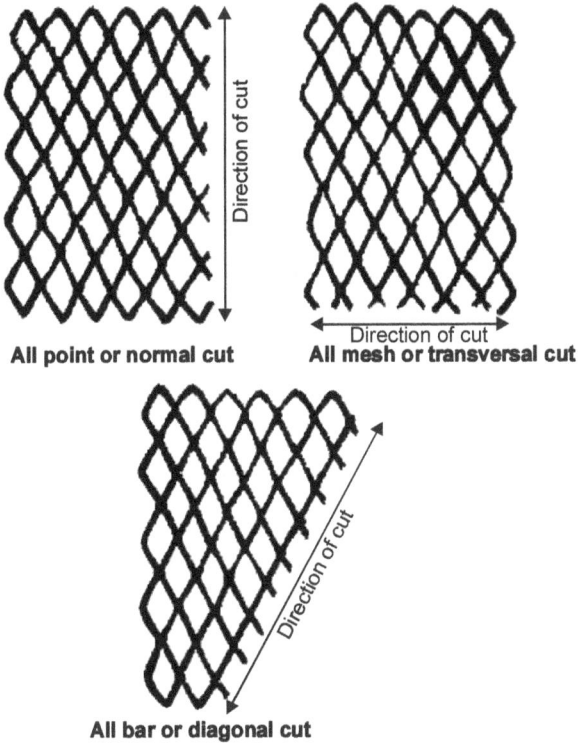

All point or normal cut   **All mesh or transversal cut**

Direction of cut

All bar or diagonal cut

Fig. 2 (a): The basic cutting patterns.

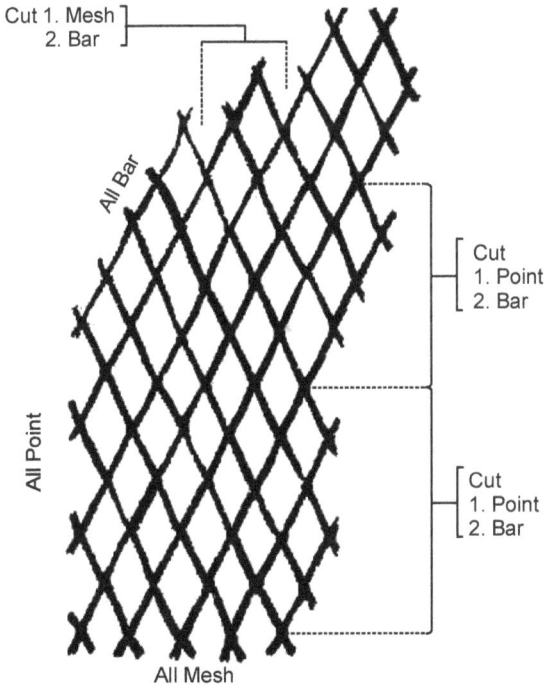

Cut 1. Mesh
     2. Bar

All Bar

All Point

Cut
1. Point
2. Bar

Cut
1. Point
2. Bar

All Mesh

Fig. 2 (b): A combination of different cutting patterns.

Point and bar cutting : When this pattern is adopted half a mesh is lost every time a bar is cut and the cut webbing become one row longer, while every time a point is cut webbing become two rows longer.

Mesh and bar cutting : Half a mesh is lost every time a bar is cut and the cut webbing become one row longer, whereas cutting a mesh does not increase the length of the cut webbing. The cutting rate or the taper ratio gives the systematical combination of different types of cuts. A netting panel shown in Fig. 2b in which different cutting patterns and their combinations are illustrated- A few principal cuts are listed below showing the degree of tapering gained/lost (Garner, 1962).

	Cut	Loss/Gain
A	All bars	1 mesh in 2 rows
B	1 point, 4 bars	1 mesh in 3 rows
C	1 point, 2 bars	1 mesh in 4 rows
D	1 point, 1 bar x2	1 mesh in 5 rows
	1 point, 2 bars	
E	1 point, 1 bar	1 mesh in 6 rows
F	1 point, 1 bar x3	1 mesh in 7 rows
G	1 point, 1 bar x3	1 mesh in 8 rows
	2 points, 1 bar	
H	2 points, 1 bar x3	1mesh in 9 rows
	1 point, 1 bar	1 mesh in 9 rows
I	2 points, 1 bar	1 mesh in 10 rows
J	3 points, 1 bar	1 mesh in 12 rows
	2 points, 1 bar	
K	5 points, 1 bar	1 mesh in 24 rows
	6 points, 1 bar	
L	All points	None
M	1 mesh, 1 bar	1 mesh in 3 mesh
N	1 mesh, 2 bars	1 mesh in 2 mesh

There is a general formula to arrive at the baiting ratio which is illustrated below;

From a rectangular webbing ABCD, a triangular panel CDE is to be cut. Applying the formula,

$$P/B = \frac{(L-N)}{2N}$$

Where P = Point; B = Bar; L = Number of meshes in length of the required panel and N = Number of meshes in height of the panel;

We get, P/B = (100-60)72x60 = 1/3; that is, IP 3B is the cutting ratio to be applied.

In hand braiding the effect of baiting can be achieved in two ways. One method, called flat baiting involves doubling the twine at the end of a row so as to cut out one mesh instead of making the mesh. (Fig. 3a (1)).

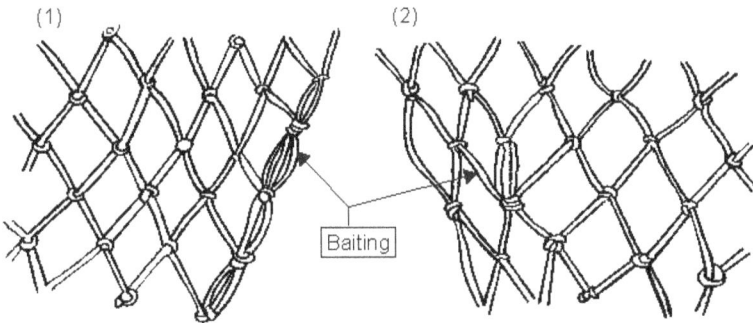

Fig. 3 (a): (1) and (2) Baiting methods.

The desired degree of tapering can be achieved by making the cutting steps at calculated intervals. Another method of reducing is to pull two meshes together when forming the next mesh below them (Fig. 3a(2).). This may be done at the ends or in the body of the net, but the latter is not advised unless a quicker taper than is possible by working at the ends is needed, as the net will be distorted.

Fishing net designs often require the number of meshes in a panel to increase and this is termed creasing. It is done by forming false meshes as shown in Fig. 3b.

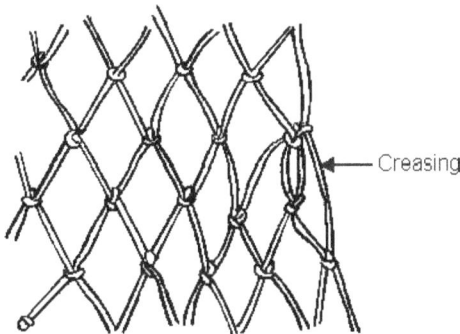

Fig. 3(b): Method of creasing.

Here again it is advisable to do it at near the end of the row to minimize distortion.

Joining of panel and take up ratio : Once the panels as per requirements are ready for a fishing net, the next step in fabrication is joining of the nettings. It is the process of connecting by means of a thread, the edge of netting panels which may differ in number of meshes, mesh size and type of cut. There are many variables while joining, but it is fairly standard practice to join together two sections with double twine or single twine with different color and greater strength. This in addition to serving as a marker will prevent ripping from one section of net to the next. The common methods of joining employed in fabrication of trawl nets are:

1. Sewing : Sewing is the process where the connecting yarn makes a knot at every mesh it passes through and in course half a mesh is formed in between the two joining edges. Some method of sewing are furnished below :

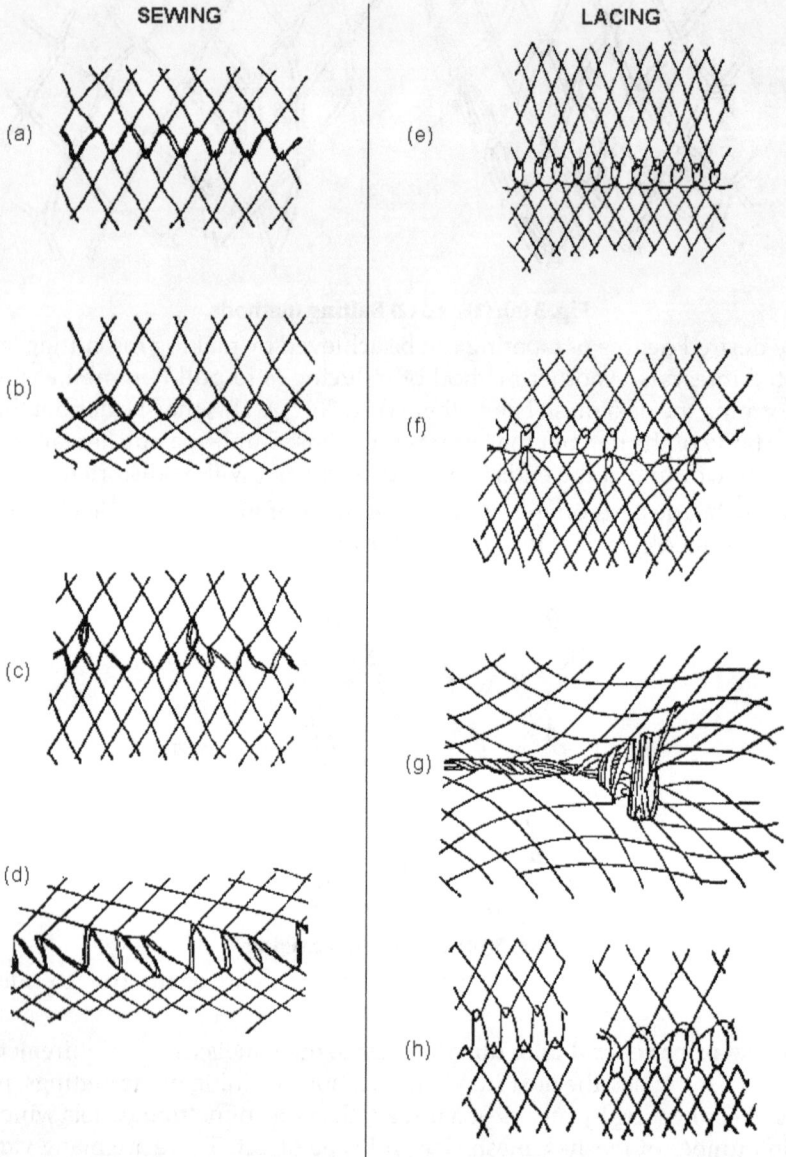

Fig. 4: Methods of joining netting panels (Sewing & Lacing)

(a)   Joining of panels with same number of meshes having same (Fig. 4 a) or different mesh sizes (Fig. 4 b).

(b)   Joining of panels with different number of meshes and either same or different mesh sizes (Fig. 4 c).

(c)   Joining of panels where the edges are cut differently (Fig. 4 d).

2. Lacing (Seaming) : Knots are not necessarily made at every mesh and no new mesh or part of mesh is formed in the process of lacing. Some commonly adopted methods are illustrated.

(a) Joining of panels with same number of meshes and either same or different mesh size (Fig. 4 e)

(b) Joining of panels with different number of meshes and either same or different mesh size (Fig. 4 f).

(c) Joining of panels with same number of meshes and same mesh size or with different number of meshes and different mesh size (Fig. 4 g).

(d) Joining of panels with same number of meshes or different number of meshes with different types of cut (Fig. 4 h).

Take up (joining ratio) : For joining of different panels, the proportion of the number of meshes on one panel's edge is indicated as a fraction or ratio. For example,

$$A/B = 3/4 \text{ or } A : B = 3 : 4$$

Three meshes of panel A to be joined with 4 meshes of panel B. Another way of expressing joining ratio is to refer to the extent of length of each panel, instead of the number of meshes. For example; A/B = 3m/4m indicates that each time, 3 meter of panel A to be joined with 4 meter of panel B.

Mounting of nettings and "Hanging coefficient" : Once the netting panels are joined together the webbing must be affixed to some kind of line or rope along its mount edge. This is described as hanging or mounting of the net. The line or rope to which the webbing is first hung before attaching to the head rope or foot rope is known as bolch line and the mounting yarn is referred as stable line. The following are some of the methods used for mounting the webbing in trawl nets.

(a) Direct Mounting

(b) Loose Mounting

(c) Mounting of Fly-Meshes

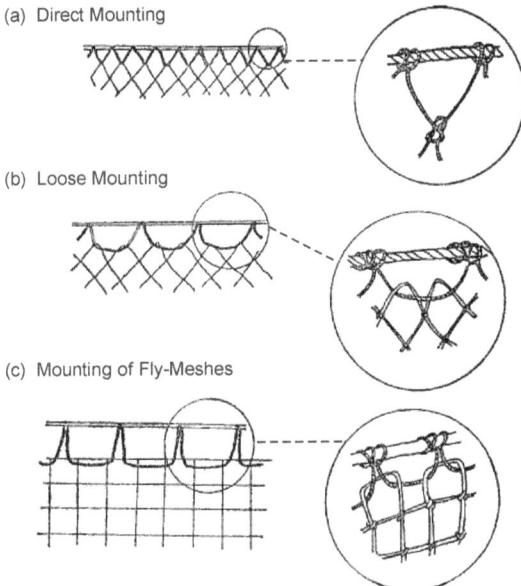

Fig. 5: Some mounting methods

1. Direct mounting or fixed mounting where each mesh is fixed directly on to the rope by means of a mounting yarn (Fig. 5a).

2. Loose mounting where the meshes are hung freely on to the mounting yarn and the mounting yarn is fixed to the rope at definite intervals (Fig. 5b).

3. Mounting of fly meshes is done by securing each mesh loosely to the rope by means of a mounting yarn tied to the rope (Fig. 5c).

The correct hanging of the webbing to the bolch line is a very important factor to get the definite degree of opening to the meshes. Variation from the desired ratio will cause strain and stress on the webbings and will result in malfunctioning as well as quick distortion of the gear. The hanging coefficient for namely, when expressed as

Al/A = 4/5 or al/A = 0.8 refers 5 meter length of webbing A to be mounted on 4 meter rope(al). In other word 1 meter webbing to be mounted on 0.8 meter rope.

On completing the hanging of webbing, the last step left in fabrication of a trawl gear is fixing of the net to the supporting ropes. This is done *by* securing the upper bolch line to the head rope and lower bolch line to the foot rope by means of sittings as illustrated in Fig 6.

Description of trawl gears operated by Exploratory Fisheries Project.

Data sheets and construction drawings are used to present maximum possible information. Figures 6-20 gives the data and construction drawings of trawl gears operated by the Exploratory Fisheries Project.

Conventions adopted in data sheets: Each data sheet consists of five sections, the first enumerates the general conditions of operation, the second specifies the webbing, the third specifies the lines and ropes, the fourth the floats and sinkers and the last section is for other vital information.

Webbing : The various netting panels are specified in capital letters which are cross-referenced with drawings.

Material: Cotton, manila, polyethylene, polypropylene are all taken as names being in common use and are used as such.

Type of knot : All nettings are made of single sheet bend knots.

Preservation : Wherever preserved, name of preservative is indicated.

Color: Whenever preserved, the color after preservation is specified. In case of untreated webbing, the original color is noted.

Twine size : Expressed in diameter taking millimeter as the unit.

Breaking strength: Indicated in Kgf (Kilogram force) for conversion to the unit Newton (N) adopted by International System of Units; 1 Kgf = 9.8 N.

Twine surface area : For the whole of webbing is given in square meter.

Stretched mesh : Mesh sizes are given in millimeters. True mesh size is used, that is, the length of one lumen plus knot with meshes stretched.

Upper edge: The number of meshes along top of each panel is given.

Lower edge : The number of meshes along bottom of each panel is taken.

Depth : The number of meshes down the side of each panel is taken.

Baiting rate : Where the panel tapers differently on its two edges, then the baiting rate for both the inner and outer edge is specified on successive lines.

Take up : The joining of different panels is indicated by a simple ratio of the number of meshes of one panel joined to the number of meshes in the other.

Quantity : The total quantity of webbing is noted in kilograms.

Hanging : The ratio of attachment of webbing of two supporting lines or rope is given.

Lines and ropes : The various lines and ropes are designated in lower case letters, using subscript numerals wherever necessary. These letters are cross-referenced with the drawings.

Material : Manila, combination rope, galvanized flexible steel wire are all taken as names being in common use and are used as such. Wherever combination rope is used the components are specified.

Construction : The number of strands are given. In case of steel wire details of inner core is also specified.

Diameter: In all cases, this is specified in millimeter.

Breaking strength: In all cases, this is specified in Kgf (Kilogram force).

Conventions adopted in drawings: The construction drawing contains the following details which also appear in data sheet

1. The number of meshes at the top and bottom of each panel.
2. The depth, in number of meshes, of each panel.
3. The mesh size of panels in millimeters.
4. The material used and twine size of the panel.
5. The fabrication detail of the panels- either single braided or double braided.
6. The baiting rates in each panel.
7. The length of ropes, section-wise, in meters.

Data and construction details of 12 meter two seam fish trawl net
Construction details

Material of webbings — Cotton twine 20Sx15x3; Type of knot — Single sheet bend; Twine size-2mm; Breaking strength- 18.5 kgf, Twine surface area— 11.100 sq.m.

Webbings	A	B	C	D	E	F	G
Stretched mesh (mm)	120	120	120	100	80	60	45
Upper edge (m)	16	3	124	96	68	50	60
Lower edge (m)	42	28	96	68	34	50	60
Depth (m)	39	67	24	26	34	25	60
Baiting rate (Inner)	1:1	1:1					
Baiting rate (Outer)	1:2	1:1.5	1:2	1:2	1:2	Nil	Nil
Take up	A/C-1/1	B/D-1/1	C/D=1/1	D/E=1/1	E/F=2/3	F/G=5/6	

Hanging	al/A=0.95; bl/B=0.95; a2/C=0.50; b2/D=0.55						
Lines and ropes	a	a1	a2	b	b1	b2	c
Material	<- Combination (& wire) -------->Manila rope						
Construction	< Cable laid (6 strands) -------->					9 strands	
Diameter (mm)	<	14-------------->				10	
Breaking strength						700	
Length (m)	2.5	4.7	2.4	0.5	7.6	2.4	12.7
Floats, Sinkers	Floats			Sinkers			
Diameter (mm)	150	130		6			
Material	Aluminum			G I chain			
Number	1	16					
Shape	Round	Round		Chain			
Weight in air (kg)				18-20			

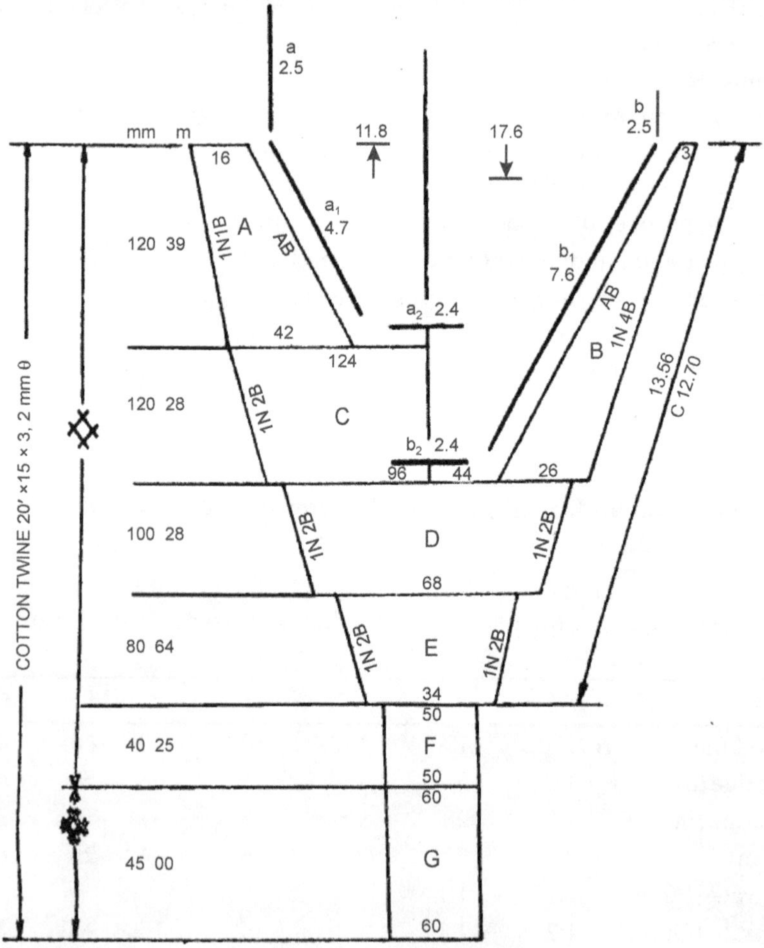

Fig. 6: Drawings of 12 m two seam fish trawl net

Head rope-11.8 m, Foot rope - 17.6 m, Bolch line - Nylon rope 5 mm dia.
Operational details

Vessels - 10.4-11.26 m OAL; GRT - 9.9-12.80; HP - 42-56; Trawling speed- 2.5 knots

Crew - 6; Otter board - Oval, 50-55 kg

Place of operation - East and West coast of India.

Data and construction details of 14 meter two seam fish trawl

Construction details

Materials of webbings - Cotton twine — 20x18x3; Type of knot — Single sheet bend, Preservation — Cutch, Twine size — 2.2 mm dia.; Breaking strength — 22.5 Kgf; Twine surface area — 19.521 sq. m; Color — Dark brown

Webbing	A	B	C	D	E	F	G
Stretched mesh (mm)	120	120	120	100	80	60	50
Upper edge (m)	23	8	174	130	92	72	72
Lower edge (m)	62	35	130	92	54	72	77
Depth (m)	46	75	32	36	38	30	90
Baiting rate (Inner)	1:1	1:1					
Baiting rate (Outer)	1:6.5	1:1.5	1:1.5	1:2	1:2		
Take up	A/C = 1/1 B/D=1/1 C/D=1/1 D/E-l/l E/F=3/4F/O=1/1						
Hanging	al/A-1.00; bl/B=0.98; a2/C=0.50; b2/D-0.50						

Lines and ropes	a	a1	a2	b	b1	b2	C
Material	<- — — Combination (Manila & wire)— — ->						Manila rope
Construction	<- — — Cable laid (6 strands) — — — — — —— ->						9 strands
Diameter (mm)	<------------ 14 --------——>						10
Breaking strength							700
Length (m)	2.5	5.5	3.0	0.5	9.8	3.0	18.2
Floats & Sinkers	Floats			Sinkers			
Diameter (mm)	200	152		9			
Material	Aluminum			G.I chain			
Number	1	14					
Shape	Round	Round					
Weight in air (Kg)				22-24			

Head rope — 14 m; Foot rope — 22 m; Bolch line — Nylon or PE rope — 6 mm

Operational details

Vessels — 14.3 m OAL; GRT — 34.5; HP — 125-165; Towing speed - 3 knots; Crews — 6, Otter board — Oval — 40-65 kg, Place of operation - East & West coast of India

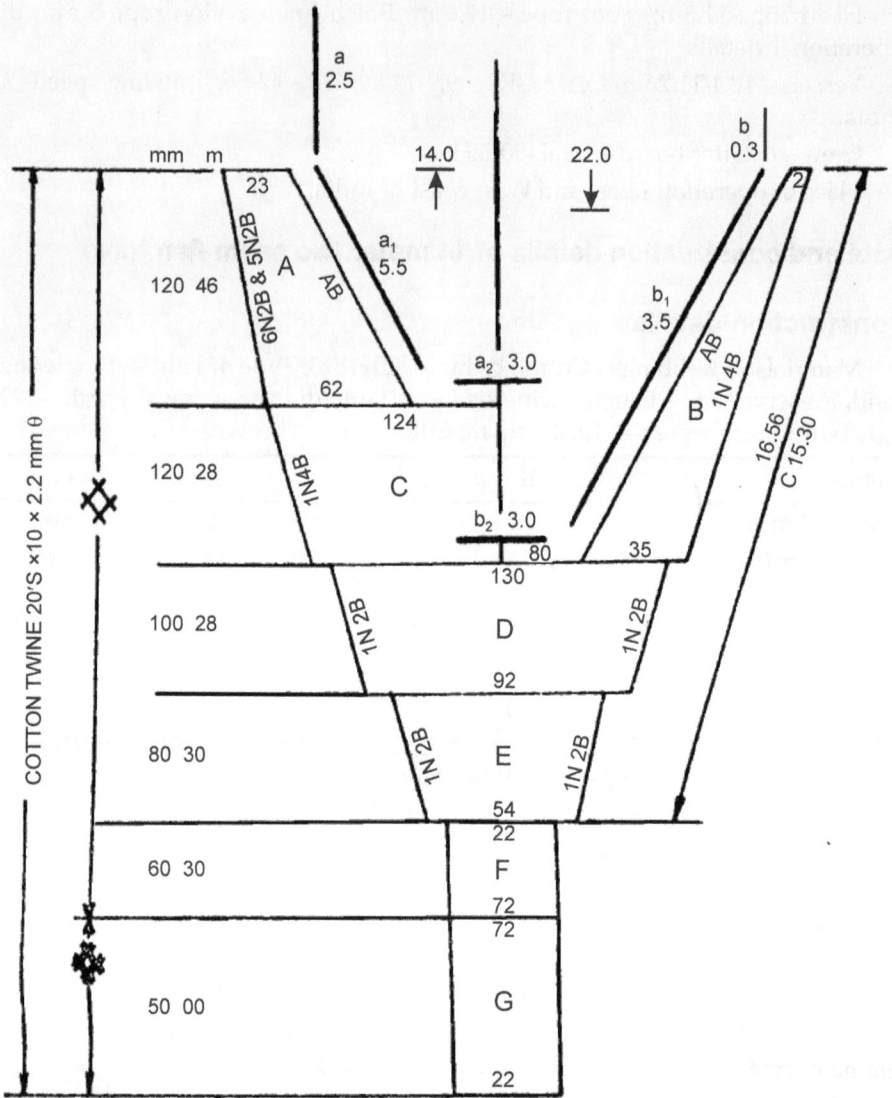

Fig. 7: Drawings of 14 m two seam fish trawl net

Data and construction details of 15 meter two seam fish trawl net

Construction details

Material of webbings — Cotton twine 20Sx20x3 or Manila twine
Type of knot—Single sheet bend
Preservation—Cutch; Color—Dark brown
Twine size (Dia.)—2.5 mm; Breaking strength (Kgf) — 25
Twine surface area (Sq. m) — 22.477

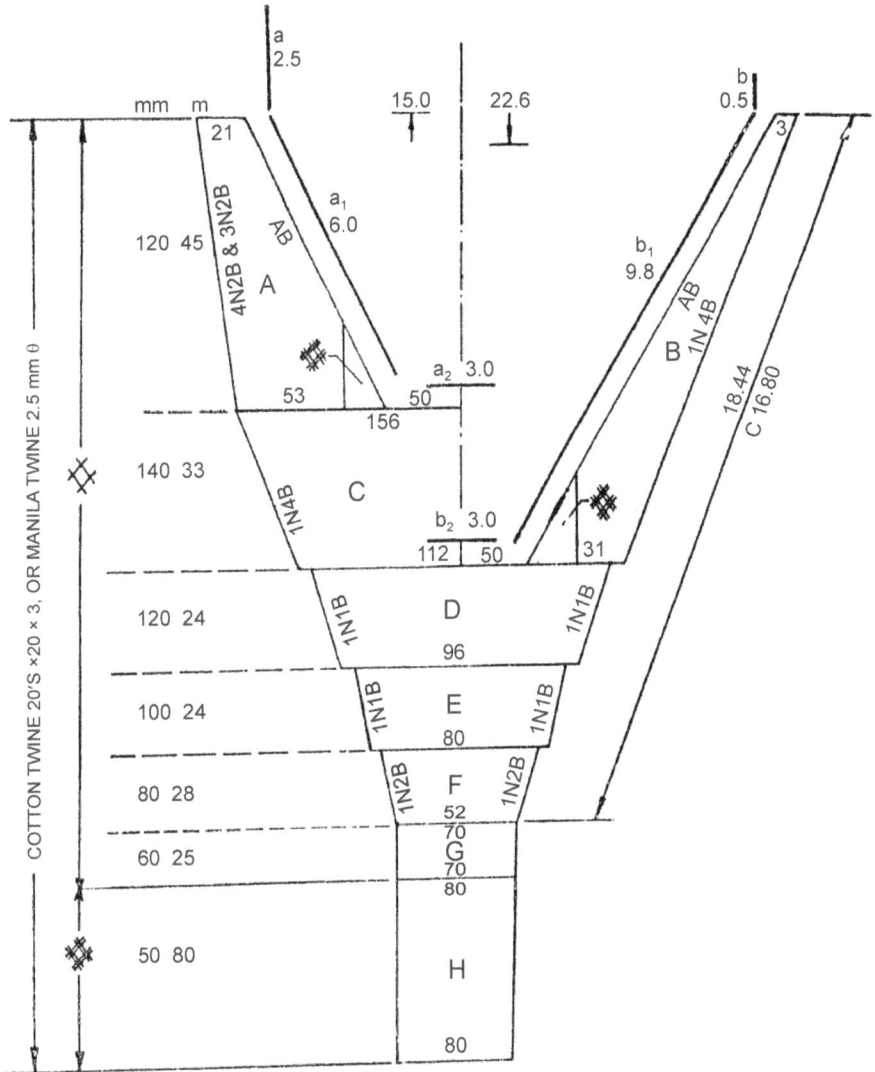

Fig. 8: Drawings of 15 meter two seam fish trawl net.

Webbings	A	B	C	D	E	F	G	H
Stretched mesh (mm)	140	140	140	120	100	80	60	50
Upperedge(m)	21	3	156	112	96	80	70	80
Lower edge (m)	53	31	112	96	80	52	70	80
Depth (m)	45	78	33	24	20	28	25	80
Baiting rate (inner)	1:1	1:1						
Bating rate (outer)	1:4.5	1:1.5	1:1.5	1:3	1:3	1:2		
Take up	A/C=1/1 B/D=1/1 C/D=1/1 D/E-1/1 E/F-1/1 F/G=5/7 G/H-7/8							
Hanging	al/A=0.95 bl/B=0.97 a2/C=0.43 b2/D=0.50							

Lines and ropes	a	al	a2	b	bl	b2	C
Material	<–Combination (Manila & wire)– – – –>Manila rope						
Construction	<–Cable laid (6 strands)– – – – – – – –9 strands						
Diameter (mm)	<– – – – – – – –20– – – – – – – –> 16						
Breaking strength (kg)							2000
Length (m)	2.5	6.0	3.0	0.5	9.8	3.0	16.8
Floats and sinkers		Floats		Sinkers			
Diameter (mm)	250	200		9			
Material	Aluminum			G.I chain			
Shape	<-Round —»						
Weight in air (kg)				32-37			

Head rope −15m; Foot rope −22.5 m; Bolch line −Nylon rope 8 mm dia.
Operational details

Vessels –26 m OAL, GRT –91.7, HP –240, Trawling speed –3 knots, Crew –14 Otter board –Oval, 180-200 kg, Place of operation - East and West coast of India.

Data and construction details of 16.5 meter two seam fish trawl net

Construction details

Material of webbings – Cotton twine 20 x20x3 or Manila twine
Type of Knot – Single sheet bend
Preservation – Cutch; Color – Dark brown
Twine size –2.5 mm diameter; Breaking strength – 25 kgf
Twine surface area in sq.m – 28.148

Webbings	A	B	C	D	E	F	G	H
Stretched mesh (mm)	140	140	140	120	100	80	60	50
Upperedge(m)	24	5	170	124	105	88	75	85
Lower edge (m)	60	33	124	106	83	56	75	85
Depth (m)	48	84	36	27	27	32	30	100
Baiting rate (inner)	1:1	1:1						
Bating rate (outer)	1:4.5	1:1.5	1:1.5	1:1.5	1:3	1:3	1:2	
Take up	A/C=1/1 B/D=1/1 C/D=1/1 D/E-1/1 E/F-1/1 F/G=3/4 G/H-15/17							
Hanging	al/A=0.57 bl/B=0.91 a2/C=0.50 b2/D=0.50							

Lines and ropes	a	al	a2	b	bl	b2	C
Material	<–Combination (Manila and wire)– – – –>Manila rope						
Construction	<–Cable laid (6 strands)– – – – – – – –(9 strands)						
Diameter (mm)	<– – – – – – –20– – – – – – – –> 16						
Breaking strength (kg)							2030
Length (m)	2.50	6.50	3.50	0.50	10.75	3.50	16.40
Floats and sinkers		Floats		Sinkers			

Diameter (mm)	250 200	9
Material	Aluminum	G.I chain
Number	20	
Shape	<-Round	—
Weight in air (kg)		33-40

Head rope –16.5 m; Foot rope – 25.0 m; Bolch line – Nylon rope 8 mm diameter
Operational details

Vessels – 26-30 m OAL; GRT - 91.7-123.2; HP –240-300; Trawling speed –3 knots Crew – 14; Otter board - Oval, 200 kg; Place of operation – East and West coast of India

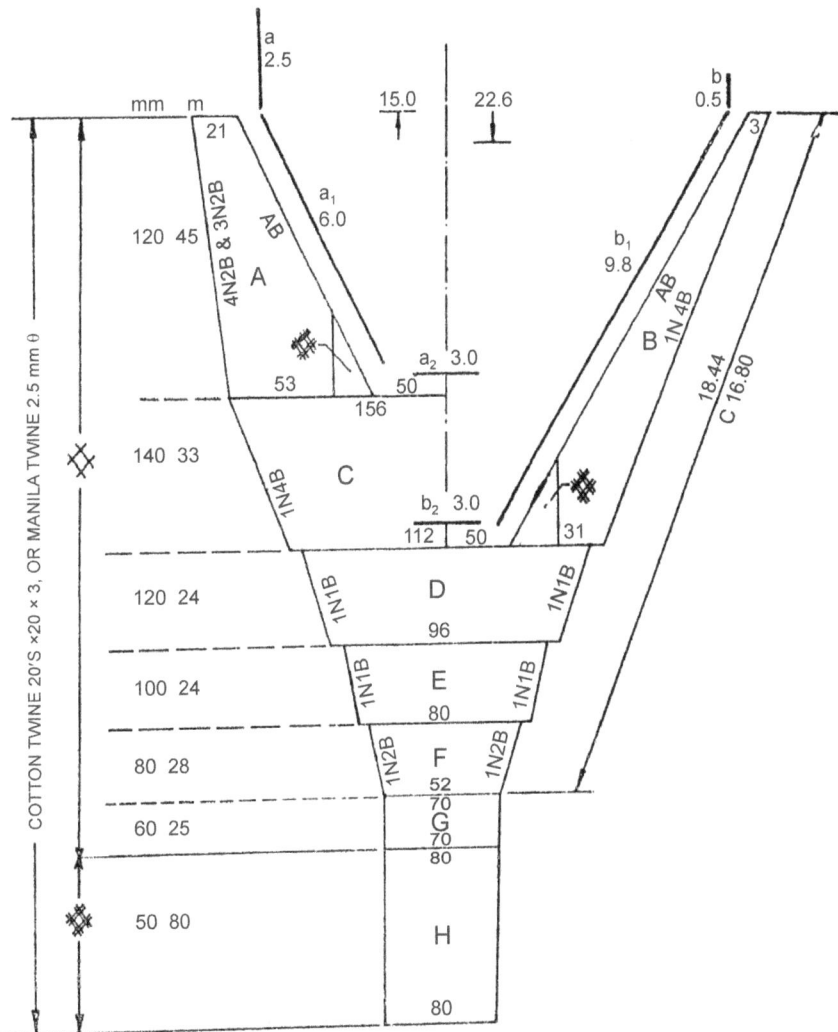

Fig. 9: Drawings of 16.5 meter two seam fish trawl net.

Data and construction details of 20 meter two seam fish trawl net

Coostruction details

Material of webbings – Polypropylene or Polyethylene

Type of knot – Single sheet bend; Color – Blue

Twine size – 2 mm diameter; Breaking strength – 36

Twine surface area – 29.035 sq.m

Webbings	A	B	C	D	E	F	G	H	I	J
Stretched mesh (mm)	140	140	120	120	120	100	60	60	50	45
Upper edge (m)	30	5	30	5	182	150	140	120	80	90
Lower edge (m)	30	5	65	41	130	112	92	66	80	90
Depth (m)	18	18	54	106	52	38	48	54	50	100
Baiting rate (Upper)	nil	nil	1:1	1:1						
Baiting rate (Lower)			1:3	1:5	1:2	1:2	1:2	1:2	nil	nil

Take up   A/C = 1/1, BD = 1/1, C/E = 1/1, DE = 1/1, E/F = 13/15, F/G = 3/4, G/H = 3/4, H/I= 5/7, I/J= 5/9

Hanging   a1/A=0.99; b1/B=0.99, a2/C=0.93, b2/D=0.94,a3/E=0.48,b3/F=0.51

Lines and ropes		a	a1	a2	a3	b	b1	b2	b3
Material	<----Galvanized flexible steel wire rope ---->								
Construction	<---6x19 with inner fiber core --->								
Diameter			9mm						
Breaking strength			100/mm square						
Length (m)	2.0	2.5	6.0	3.0	0.5	2.5	12.0	3.5	
Floats and sinkers		Floats		Sinkers					
Diameter (mm)		200		90					
Material		Plastic		G.I chain					
Number		10-12							
Shape		Round							
Weight in air (kg)				30-35					

Head rope – 20m; Foot rope – 32.5m; Bolch line – Nylon rope 8 mm diameter

Operational details

Vessels – 16.8-17.5 m OAL; GRT – 48.7-56.8; HP – 153-210; Trawling speed –3 knots Crew –7–10, Otter board – Oval 100 kg; Place of operation-East & West coast of India

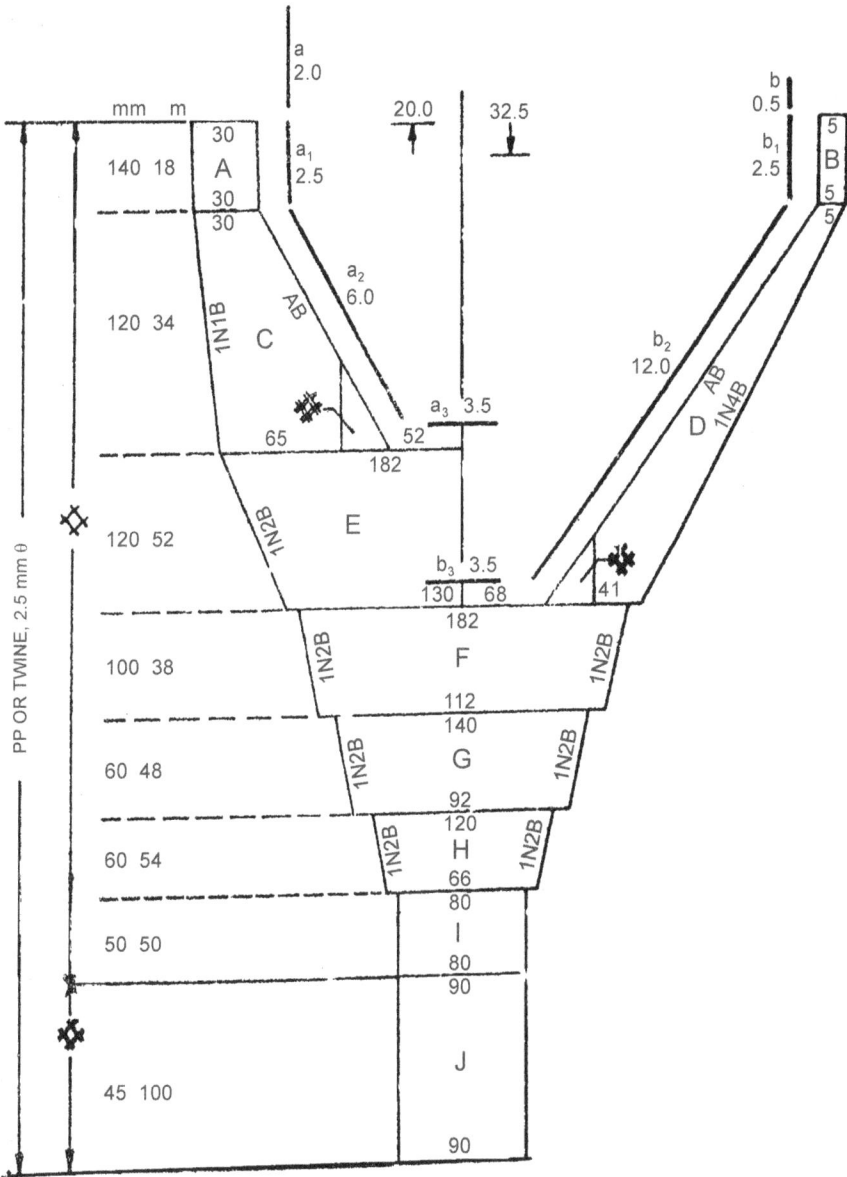

Fig. 10 : Drawings of 20 meter two seam fish trawl net

Data and construction details of 24 meter two seam fish trawl net

Construction details

Material of webbing – PP or PE

Type of knot – Single sheet bend; Color – Blue; Twine size – 2 mm diameter

Breaking strength – 36; Twine surface area – 30.275 sq. m

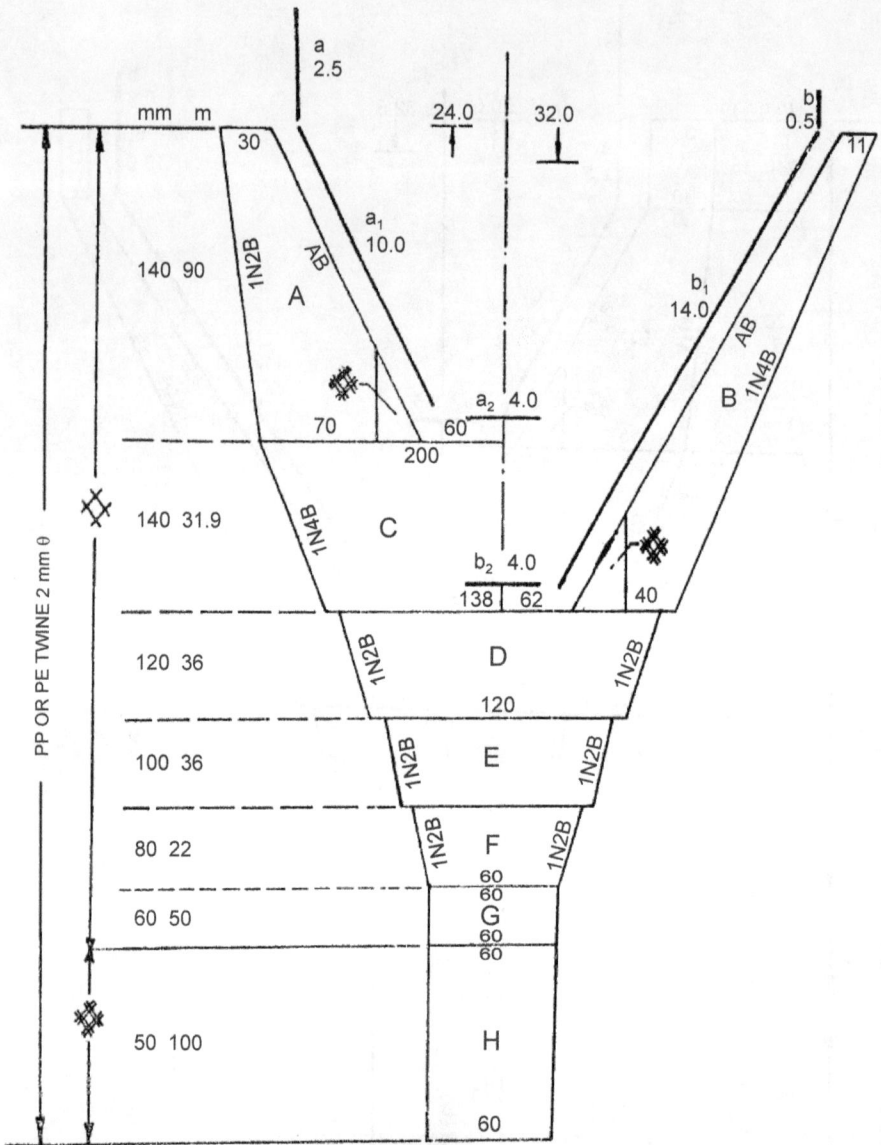

Fig. 11: Drawings of 24 meter two seam fish trawl net

Webbings	A	B	C	D	E	F	G	H
Stretched mesh (mm)	←	140	→	120	100	80	60	50
Upper edge(m)	30	11	200	150	120	82	80	80
Lower edge (m)	70	48	158	120	82	60	80	80
Depth (m)	80	111	31	36	36	22	50	100
Baiting rate (inner)	1:1	1:1						
Bating rate (outer)	1:2	1:1.5	1:1.5	1:2	1:2	1:2		

Take up	A/C=1/1 B/D=1/1 C/D=1/1 D/E-1/1 E/F-1/1 F/G=3/4 G/H-1/1					
Hanging	al/A=0.89 bl/B=0.89 a2/C=0.48 b2/D=0.46					
Lines and ropes	a	al	a2	b	bl	b2
Material	<—Galvanized flexible steel wire rope ————>					
Construction	<—6x24 or 6x19 with inner fiber core ———————>					
Diameter (mm)	<——————11—————————>					
Breaking strength (kg)	180/sq. mm					
Length (m)	2.5	10	4	0.5	14	4

Floats and sinkers	Floats		Sinkers
Diameter (mm)	250	200	9
Material	Aluminum		G.I chain
Shape	Round		
Weight in air (kg)	45-50		

Head rope – 24 meter; Foot rope – 32 meter; Bolch line – Nylon or PE rope 8 mm dia. Operational details

Vessels – 16.8 and 17.5 OAL; GRT – 49-57; HP – 153-210; Trawling speed – 3 knots Crew – 7-10, Otter board – Oval, 180 kg, Place of operation – East and West coast of India

Data and construction details of 24 meter modified two seam fish trawl net

Construction details

Material of webbing – PP or PE
Type of knot – Single sheet bend; Color – Blue; Twine size – 2 mm diameter
Breaking strength – 38

Webbings	A	B	C	D	E	F	G	H
Stretched mesh (mm)	<——70——>		60	50	<——40——>			
Upperedge(m)	60	22	400	315	240	164	100	100
Lower edge (m)	140	96	216	240	164	130	100	100
Depth (m)	160	223	63	72	72	44	70	120
Baiting rate (inner)	1:1	1:1						
Bating rate (outer)	1:2	1:1.5	1:1.5	1:2	1:2	1:2		
Take up	A/C=1/1 B/D=1/1 C/D=1/1 D/E-1/1 E/F-1/1 F/G=5/5 G/H-1/1							
Hanging	al/A=0.89 bl/B=0.89 a2/C=0.46 b2/D=0.46							
Lines and ropes	a	al	a2	b	bl	b2		
Material	<—Galvanized flexible steel wire rope --->							
Construction	6x24 or 6x19 with inner fiber core							
Diameter (mm)	11mm							
Breaking strength (kg)	180/sq. mm							

Length (m)	2.5	10	4	2.5	14	4
Floats and sinkers		Floats		Sinkers		
Diameter (mm)	250	200		9		
Material	Aluminum			G.I chain		
Number	1	14-15				
Shape	Round					
Weight in air (kg)				50-55		

Head rope – 24 m; Foot rope – 32 m; Bolch line – Nylon or PE rope 8 mm diameter

Operational details

Vessels – 17.5 m OAL, GRT – 57, Engine HP – 200, Trawling speed – 2.5 knots
Crew – 10, Otter board – Oval – 100 kg, Place of operation – East coast of India

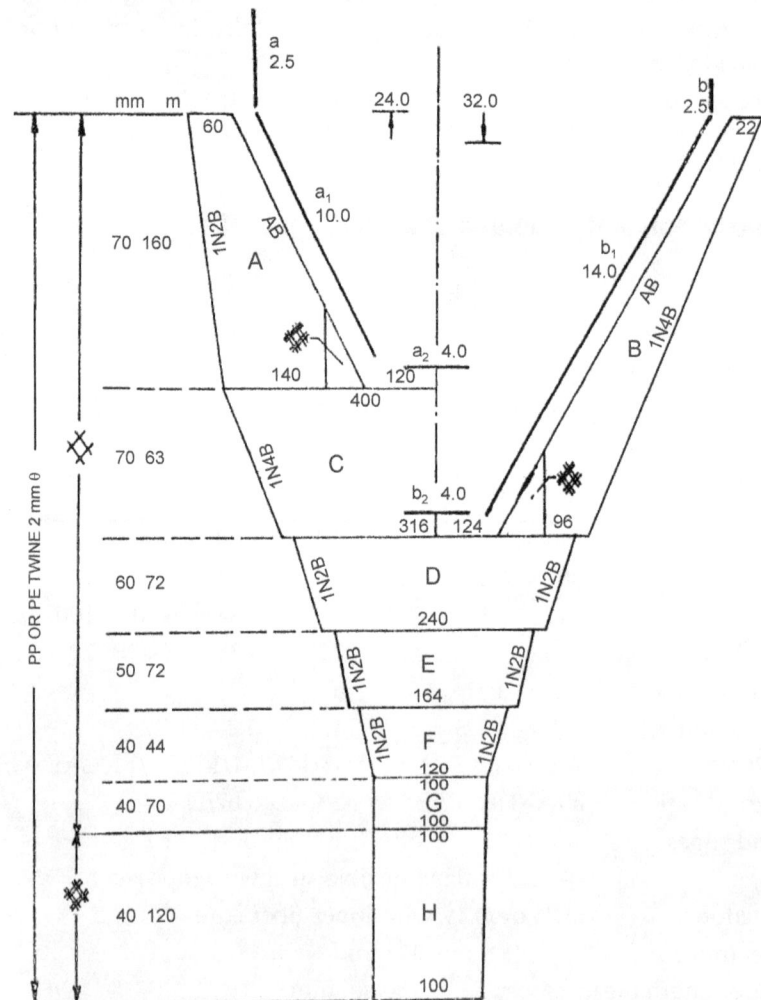

Fig. 12: Drawings of 24 meter modifed two seam fish trawl net.

Data and construction details of 30 meter two seam fish trawl net

Construction details

Webbings	A	B	C	D	E	F	G	H	I	J
Stretched mesh (mm)	←		140		→	120	100	80	60	50
Upper edge (m)	30	11	30	11	200	158	120	82	80	80
Lower edge (m)	30	11	70	48	158	120	82	60	80	80
Depth (m)	25	25	80	111.5	31.5	36	36	22	50	100
Baiting rate (Upper)	nil	nil	1:1	1:1						
Baiting rate (Lower)			1:2	1:1.5	1:1.5	1:2	1:2	1:2		

Take up	A/C = 1/1, B/D = 1/1, C/E = 1/1, D/F = 1/1, E/F = 1/1, F/G = 1/1, G/H = 1/1, H/I = 3/4, I/J= 1/1
Hanging	a1/A=0.86; b1/B=0.86, a2/C=0.89, b2/D=0.89, a3/E=0.48, b3/F=0.46

Lines and ropes	a	a1	a2	a3	b	b1	b2	b3
Material	Galvanized flexible wire							
Construction	6x19 or 6x24 with inner fiber core							
Diameter	11mm							
Breaking strength	180/ sq. mm							
Length (m)	2.5	3	10	4	0.5	3	14	4

Floats and sinkers	Floats		Sinkers	
Diameter (mm)	250	200	9	
Material	Aluminum		G.I chain	
Number	1	14-16		
Shape	Round			
Weight in air (kg)		50-55 Kg		

Head rope - 30 m; Foot rope - 38 m; Bolch line - Nylon or PE rope 8 mm diameter
Operational details

Vessels – 22-31 m OAL, GRT – 69-123, HP – 262-300, Trawling speed – 3 knots
Crew – 13-14; Otter board – Oval, 200 kg

Place of operation – East and West coast of India

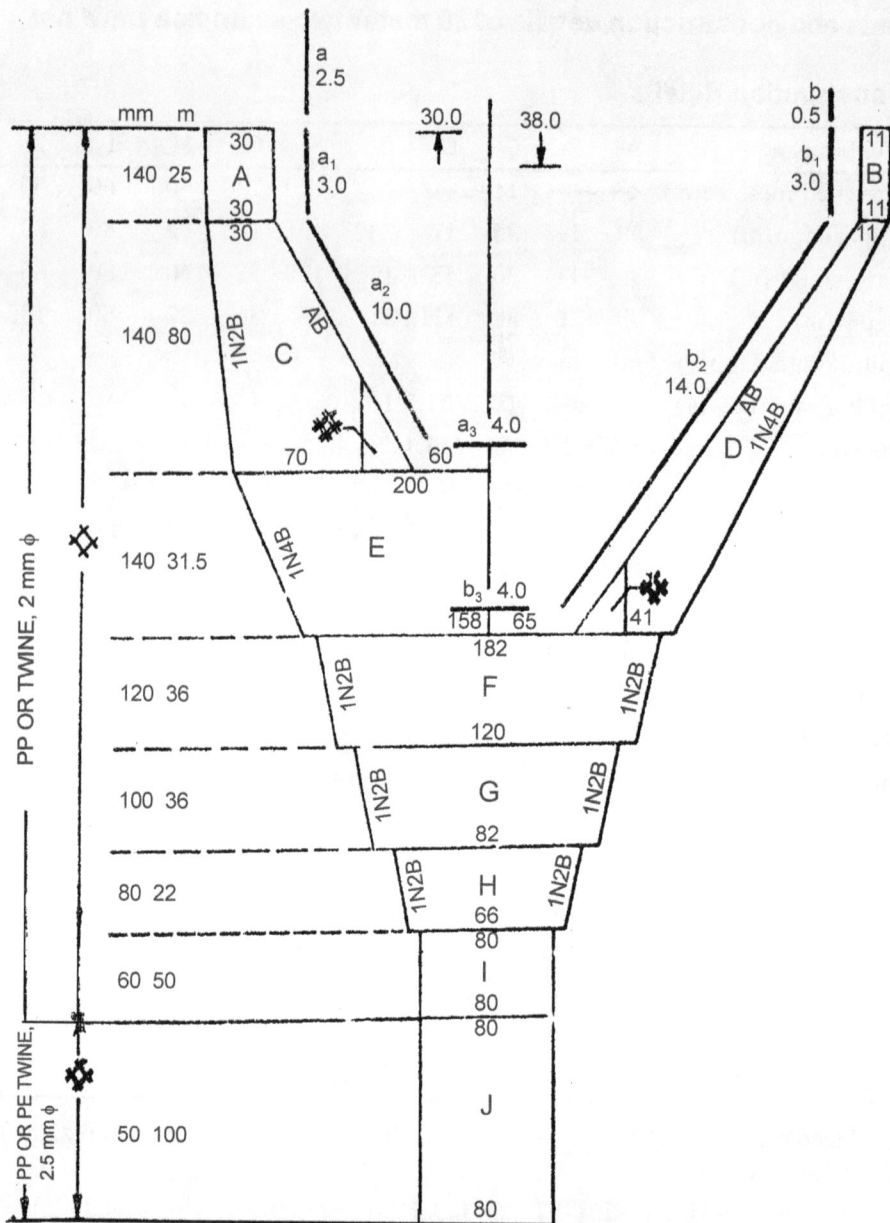

Fig. 13: Drawings of 30 meter two seam fish trawl net

Data and construction details of 35 meter two seam fish trawl net

Construction details

Material of webbing–PP or PE

Type of knot – Single sheet bend, Color – Blue, Twine size –2.5 mm diameter

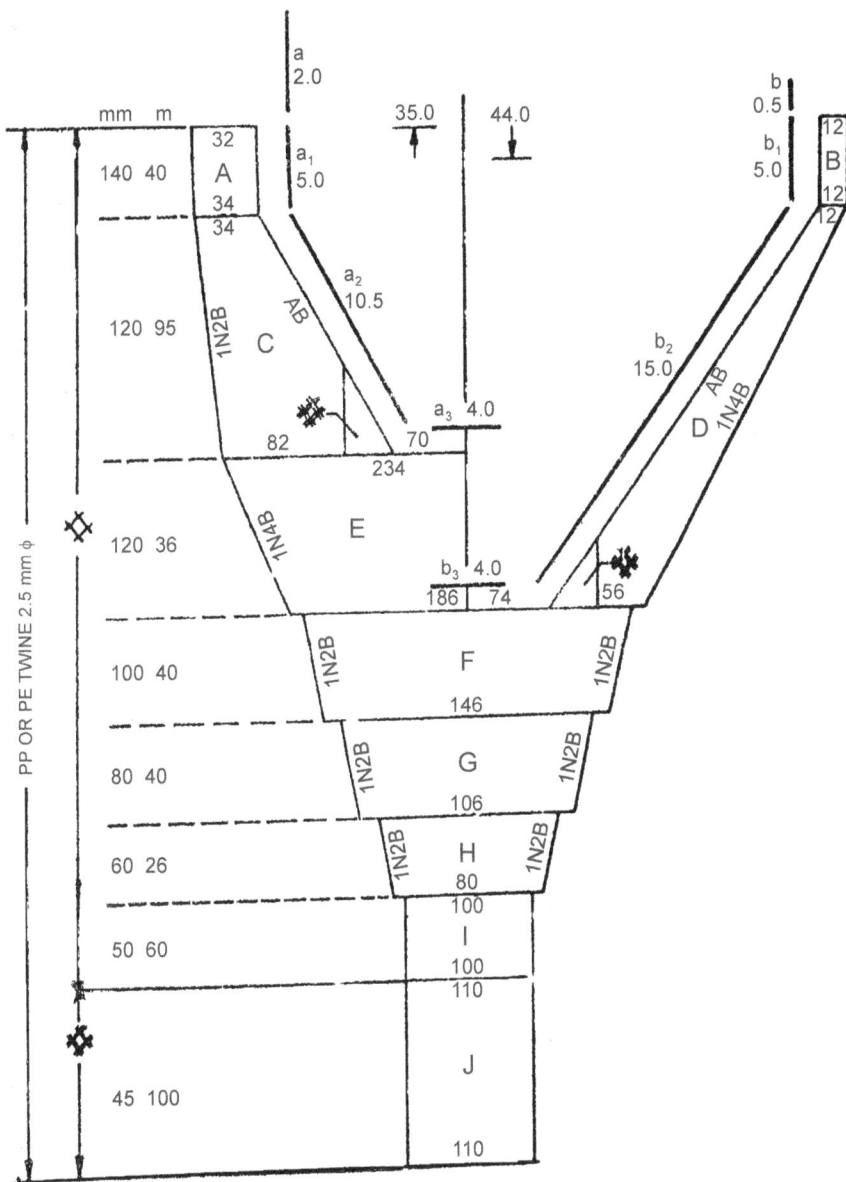

Fig. 14: Drawings of 35 meter two seam fish trawl net

Breaking strength – 63 kg, Twine surface area – 42.09 sq. m

Webbings	A	B	C	D	E	F	G	H	I	J
Stretched mesh (mm)	140	140	120	120	120	100	80	60	50	45
Upper edge (m)	34	12	34	12	234	186	146	106	100	110
Lower edge (m)	34	12	82	56	186	146	106	80	100	110
Depth (m)	40	40	95	131	36	40	40	26	60	100
Baiting rate (inner)	nil	nil	1:1	1:1						

Baiting rate (outer)	1:2 1:1.5 1:1.5 1:2 1:2 1:2 nil nil
Take up	A/C = 1/1, B/D = 1/ 1, C/E = 1/1, D/F = 1/1, E/F = 1/1, F/G = 1/1, G/H = 1/1, H/I= 4/5, I/J= 10/11
Hanging	a1/A=0.89; b1/B=0.89, a2/C=0.92, b2/D=0.95,a3/E=0.48,b3/F=0.54

Lines and ropes	a a1 a2 a3 b b1 b2 b3
Material	<-----Galvanized flexible steel wire rope ---->
Construction	<---6x19 with inner fiber core--->
Diameter	11mm;
Breaking strength	180/mm square
Length (m)	2.0 5.0 10.5 4.0 0.5 5.0 15.0 4.0
Floats and sinkers	Floats Sinkers
Diameter (mm)	200 9
Material	Plastic G.I chain
Number	16-18
Shape	Round
Weight in air (kg)	55-60

Head rope –35.0m; Foot rope – 44 m, Bolch line Nylon rope 8 mm diameter
Operational details

Vessels – 22.5-30.96 OAL, GRT – 69.2-123.2, HP – 262-300, Trawling speed – 3 knots, Crews – 13-14, Otter board – Oval 250 kg, Place of operation – East and West coast of India

Data and construction details of 45 meter two seam fish trawl net

Construction details

Material of webbing – PP or PE ; Type of knot – Single sheet bend, Color – Blue Twine size – 2.5 mm diameter; Breaking strength – 53 kg

Twine surface area – 76.85 sq.m

Webbings	A	B	C	D	E	F	G	H	I	J
Stretched mesh (mm)	140	140	140	140	140	120	100	80	60	50
Upper edge (m)	45	15	45	15	300	234	160	130	100	100
Lower edge (m)	45	15	105	22	234	180	130	80	100	100
Depth (m)	38	38	120	170	50	34	30	50	50	100
Baiting rate (inner)	nil	nil	1 : 1	1 : 1						
Baiting rate (outer)			1:2	1:1.5	1:1.5	1:2	1:2	1:2	nil	nil
Take up	A/C = 1/1, B/D = 1/ 1, C/E = 1/1, D/F = 1/1, E/F = 1/1, F/G = 1/1, G/H = 1/1, H/I= 4/5, I/J= 1/1									
Hanging	a1/A=0.86; b1/B=0.86, a2/C=0.89, b2/D=0.90,a3/E=0.40,b3/F=0.56									

Lines and ropes	a	a1	a2	a3	b	b1	b2	b3
Material	Galvanized flexible steel wire rope							
Construction	6x19 with inner fiber core							
Diameter	11mm							
Breaking strength	160/mm square							
Length (m)	2.5	4.5	15.0	6.0	0.5	4.5	21.5	6.0
Floats and sinkers	Floats	Sinkers						
Diameter (mm)	200	9						
Material	Aluminum G.I chain							
Number	18-22							
Shape	Round							
Weight in air (kg)	70-80 kg							

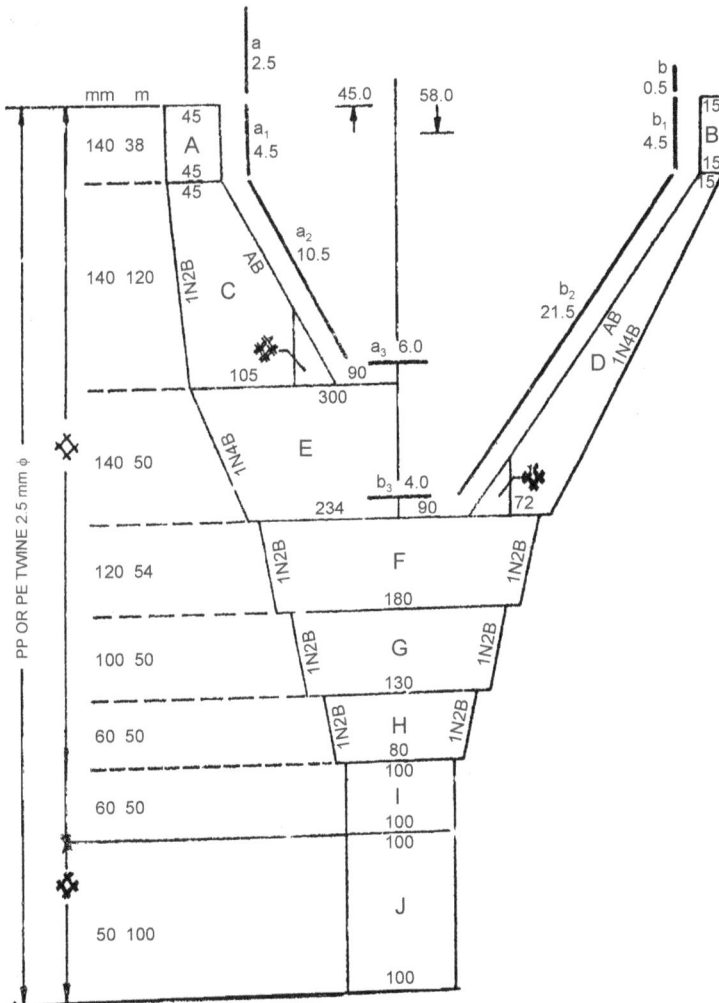

Fig. 15: Drawings of 45 meter two seam fish trawl net

Head rope –45.0m; Foot rope –55m; Bolch line – Nylon or PE rope 8 mm diameter Operational characteristics

Vessels – 30.56–38.28 m OAL, GRT – 123.2–182.6; Trawling speed– 3–3.5 knots

Crew– 14–16

Otter board – Rectangular, 425 kg

Place of operation – Upper east coast and north west coast of India.

Data and construction details of 17 meter four seam shrimp trawl net

Construction details

Material – Cotton twine; Type of knot – Single sheet bend; Preservation — Cutch Color — Dark brown

Webbings	A	B	C	D	E	F	G	H	I	J
Twine size (dia)	<—1.1 mm —>				5 mm	<—1.3mm—>			2 mm	
Breaking strength (kg)	11.1				24.0		14.1		22.5	
Twine surface area	16.97 sq. m									
Stretched mesh (mm)	76	51	51	51	51	102	51	38	38	32
Upper edge (m)	45	50	80	80	1	60	260	160	40	66
Lower edge (m)	40	50	80	34	70	60	120	54	2	64
Depth (m)	25	50	70	140	70	12	140	80	80	150
Baiting rate (inner)	nil	1:6		1:6	1:1	nil	1:4	1:1.5	1:4	nil
Baiting rate (outer)		1:6	nil	nil	nil	nil	1:2		1:4	
Take up	A/B = 2/3, B/C = 1/1, C/D = 1/1, E/G = 1/1, F/G = 1/2, G/H = 3/4, H/I = 5/6									
Lines and ropes	a	a1	a2	a3	a4	b	b1	b2	b3	b4
Material	Manila rope									
Construction	Cable laid (Total 9 strands)									
Diameter (mm)	<—16 —>						<—24—>			
Breaking strength (kg)	1600						2500			
Length (m)	1.0	1.4	2.4	3.2	3.2	1.0	1.4	2.4	3.2	3.2
Floats and sinkers	Floats		Sinkers							
Diameter (mm)	127									
Material	Aluminum		Lead							
Number	15-17									
Shape	Oval		Barrel							
Weight in air (kg)	20.22									

Head rope – 17.2 m; Foot rope – 17.2 m; Bolch line – Cotton rope 5 mm diameter Operational details

Vessels – 10.4 m OAL, GRT – 9.9-12.8, HP – 42-58; Trawling speed – 2.5 knots Crew-6; Otterboard – Rectangular 55 kg; Place of operation – West coast of India

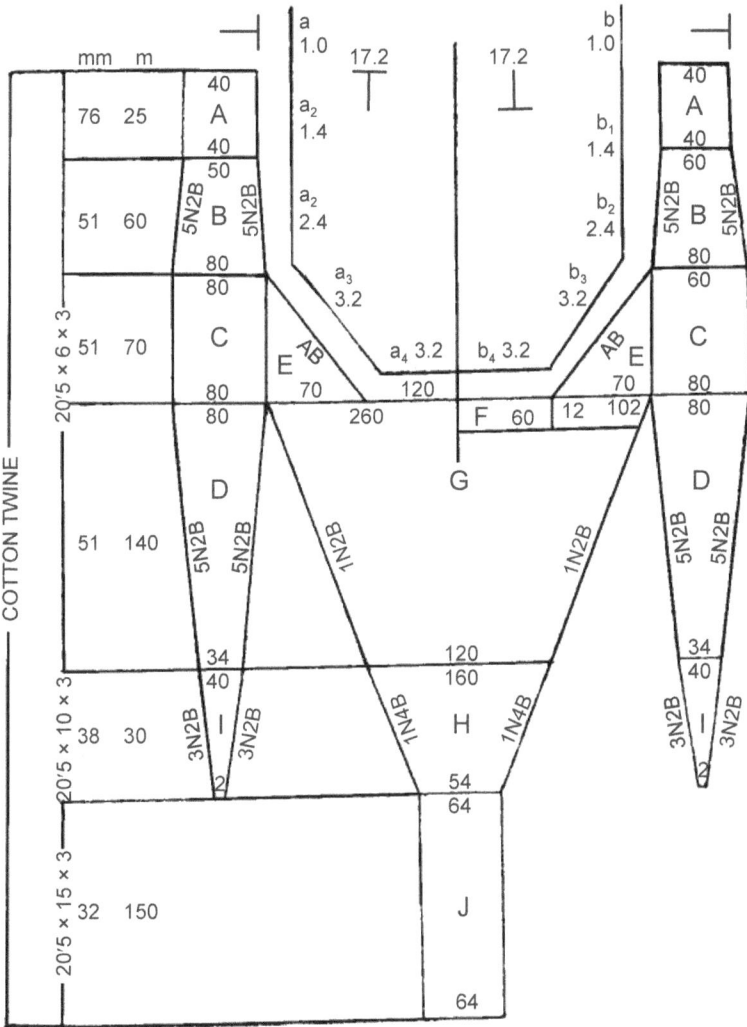

Fig. 16: Drawings of 17 meter four seam shrimp trawl net

Data and construction details of 18 meter four seam shrimp trawl net

Construction details

Material – Cotton twine, 20Sx8x3,20Sx10x3,20Sx15x3 and 20Sx20x3

Type of knot – Single sheet bend

Preservation – Cutch; Color – Dark brown

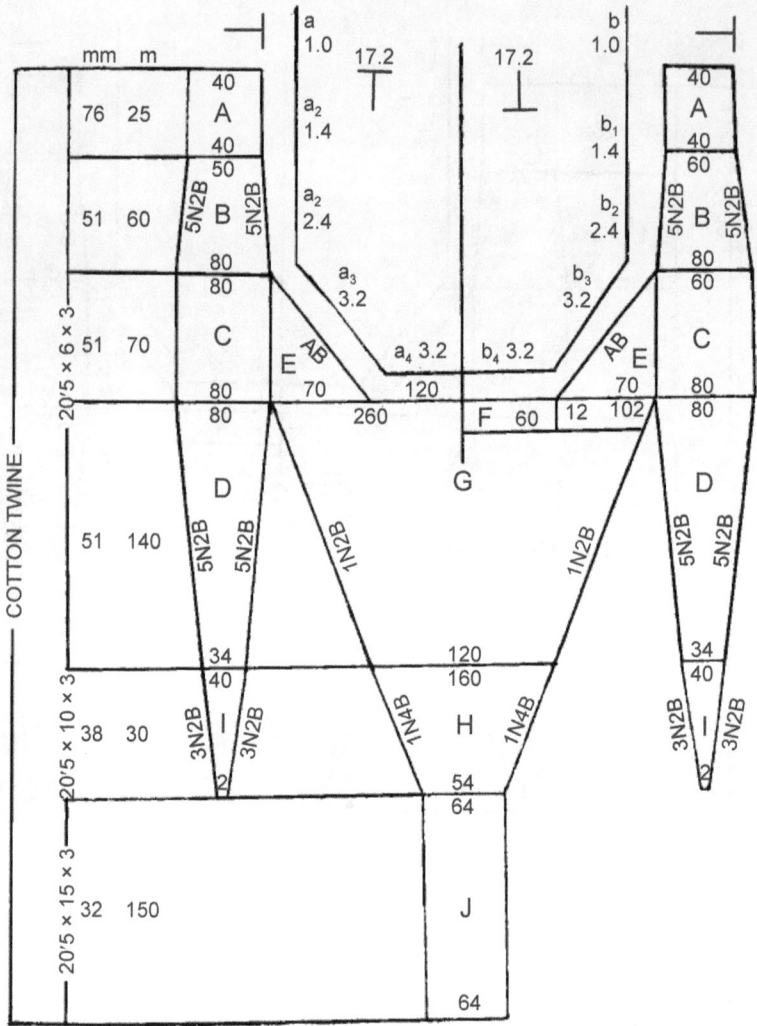

Fig. 17: Drawings of 18 meter four seam shrimp trawl net

Webbing	A	B	C	D	E	F	G	H	I	J
Twine size (dia)	<------------1.1 mm--->					<-2.5 mm		<-1.3 mm		2 mm
Breaking strength (kg)		11.2					24.6		16.1		22.6
Twine surface area in sq. m			<--------------20.31 ------------------>								
Stretched mesh (mm)	70	51	51	51	51	102	51	36	38	32
Upper edge (m)	40	60	80	30	1	60	280	150	50	80
Lower edge (m)	40	80	80	40	80	60	120	70	4	60
Depth (m)	30	60	80	150	80	17	160	80	80	160
Baiting rate (inner)	0	1:5	nil		1:1	nil				nil
Baiting rate (outer)		1:6		1:8				1:2	1:2	1:3.5

Take up	A/B = 2/3; B/C=1/1; C/D = 1/1, D/I = 4/3; E/G = 1/1 F/G = 1/2; G/H= 4/5, H/J =7/5									
Hanging	a1/A = 0.70, a2/B= 0.75, a3/C=0.68;b2/F=0.52;a2/G=0.5									
Lines and ropes	a	a1	a2	a3	a4	b	b1	b2	b3	b4
Material	Manila rope									
Construction	Cable laid (total 9 Strands)									
Diameter (mm)	<-----------15----------->				<--------------------24---------->					
Breaking strength (kg)	1800				2500					
Length (m)	1.0	1.6	2.4	3.6	3.2	1.0	1.6	2.4	3.6	3.2
Flats and sinkers	Floats		Sinkers							
Diameter (mm)	127									
Material	Aluminum		Lead							
Number	18-19									
Shape	Round	Barrel								
Weight in air (kg)	23-25									

Head rope – 18.4 m; Foot rope – 18.4 m; Bolch line – Cotton rope 5mm diameter

Operational details

Vessels – 11.20-13.70 m LOA; GRT – 12.8-15, HP – 55-60; Trawling speed – 2-2.5 knot; Crew– 6-7; Otter board – Rectangular 55-60 kg; Place of operation – West coast of India.

Data and construction details of 27 meter four seam shrimp trawl net

Construction details

Webbing	A	B	C	D	E	F	G	H	I
Material	<---------------PP or PE ------------------------------>								
Type of knot- Single sheet bend, Twine color- Blue									
Twine size (dia)	<----------------------------1.5 mm ------------> 2.5 mm								
Breaking strength (kg)	24							63	
Twine surface area (sq. m)	<----------------------45.95 ---------------- >								
Stretched mesh (mm)	64	51	51	51	38	51	51	38	32
Upper edge (m)	60	81	112	112	70	1	375	210	100
Lower edge (m)	60	112	112	56	10	112	163	90	100
Depth (m)	30	125	112	212	120	112	212	120	180
Baiting rate (inner)	1:2				1 : 1				
Baiting rate (outer)	nil	nil	nil	1:75	1:4	nil	1:2	1:2	nil
Take up	A/B=3/4, B/C=1/1; C/D=1/1; D/E=4/5, F/G=1/1; G/H=3/4; H/I=9/10								
Hanging	a1/A=0.94 a2/B=0.74								
Lines and ropes	a a1 a2 a3 a4 b b1 b2 b3 b4								
Material	Galvanized flexible steel wire rope								

Construction	6x19 with inner fiber core
Diameter	9 mm
Breaking strength (kg)	160/mm square

| Length (m) | 1.0 | 1.8 | 5.0 | 4.5 | 4.5 | 1.0 | 1.8 | 5.0 | 4.5 | 4.5 |

Floats and sinkers	Floats	Sinkers
Diameter (mm)	200	9
Material	Plastic	GI chain
Number	9-10	
Shape	Round	
Weight in air (kg)		33-38

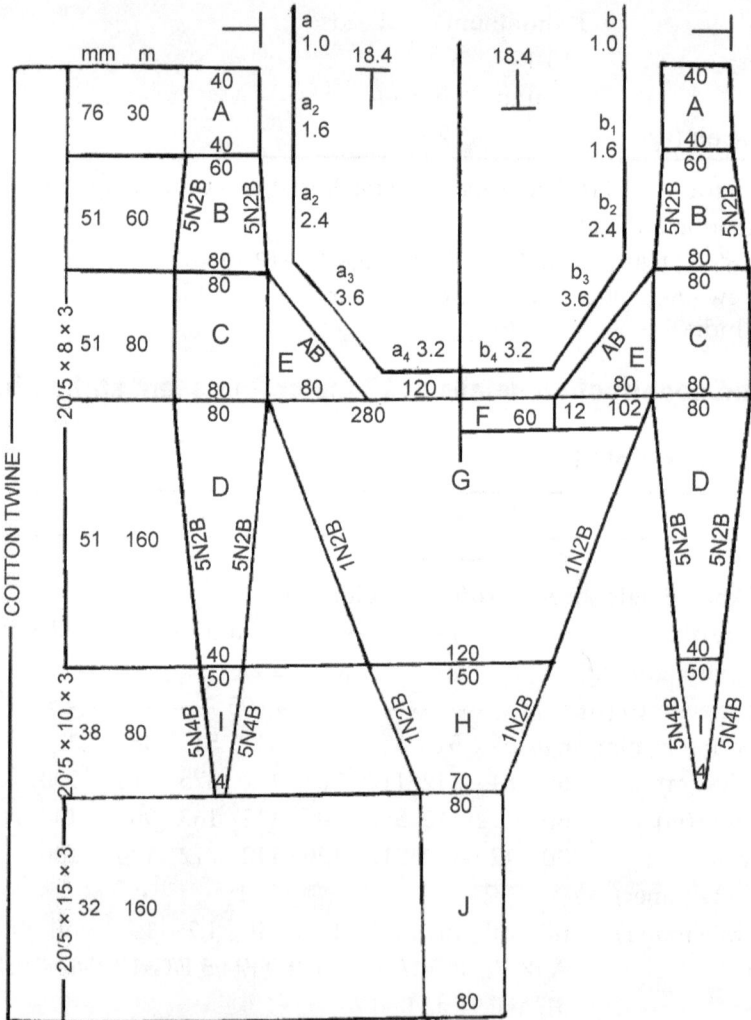

Fig. 18: Drawings of 27 meter four seam shrimp trawl net

Head rope – 27.1 m; Foot rope – 27.1 m, Bolch line – PE, 8 mm diameter
Operational details
Vessels – 17.5 m OAL, GRT– 56.8; HP – 210, Trawling speed – 2.5 knots; Crew – 10
Otter board –Rectangular, 120 kg or Oval–150 kg
Place of operation – East anf West coast of India

Data and construction details of 28 meter four seam shrimp trawl net

Construction details

Material – PP or PE, Type of knot – Single sheet bend, Color of twine – Blue

Webbing	A	B	C	D	E	F	G	H	I	J
Twine size	<----------------------1.5 mm------------------->								2.5mm	
Breaking strength (kg)					24					63
Twine surface area					17.16 sq. m					
Stretched mesh (mm)	50	50	50	50	50	50	50	50	40	36
Upper edge (m)	44	74	74	1	440	1	60	360	90	100
Lower edge (m)	74	74	6	120	80	60	80	80	90	100
Depth (m)	150	120	340	120	270	60	60	210	80	140
Baiting rate (inner)	1:5		1:5	1:1		1:1	1:1		nil	nil
Baiting rate (outer)		nil			1:1.5		1:1.5			
Take up	A/B = 1/1; B/C=1/1; D/E = 1/1, E/I = 8/9; H/I = 8/9 I/J = 9/10;									
Hanging	a1/A = 0.71, b1/A=0.71 b2/D=0.73 a2/D=0.92 a3/E=0.65 b3/F=0.92 B4/H=0.65									
Lines and ropes	a	a1	a2	a3		b	b1	b2	b3	b4
Material	Galvanized flexible steel wire rope									
Construction	6x19 with inner fiber core									
Diameter (mm)				9 mm						
Breaking strength (kg)				160/sq.mm						
Length (m)	1.0	5.3	5.3	6.5		1.0	5.3	2.2	5.5	6.5
Floats and sinkers		Floats		Sinkers						
Diameter (mm)		200		9						
Material		Plastic		GI chain						
Number		11-12								
Shape		Round								
Weight in air (kg)				45-50						

Head rope – 28.1 m, Foot rope – 32.5 m, Bolch line – PE, 6 mm diameter

Operational details

Vessels – 17.5 m OAL, GRT – 56.8, HP – 210, Trawling speed – 2.5 knots, Crew–10, Otter board – Rectangular 120 kg or Oval – 150 kg, Place of operation – Indian coasts

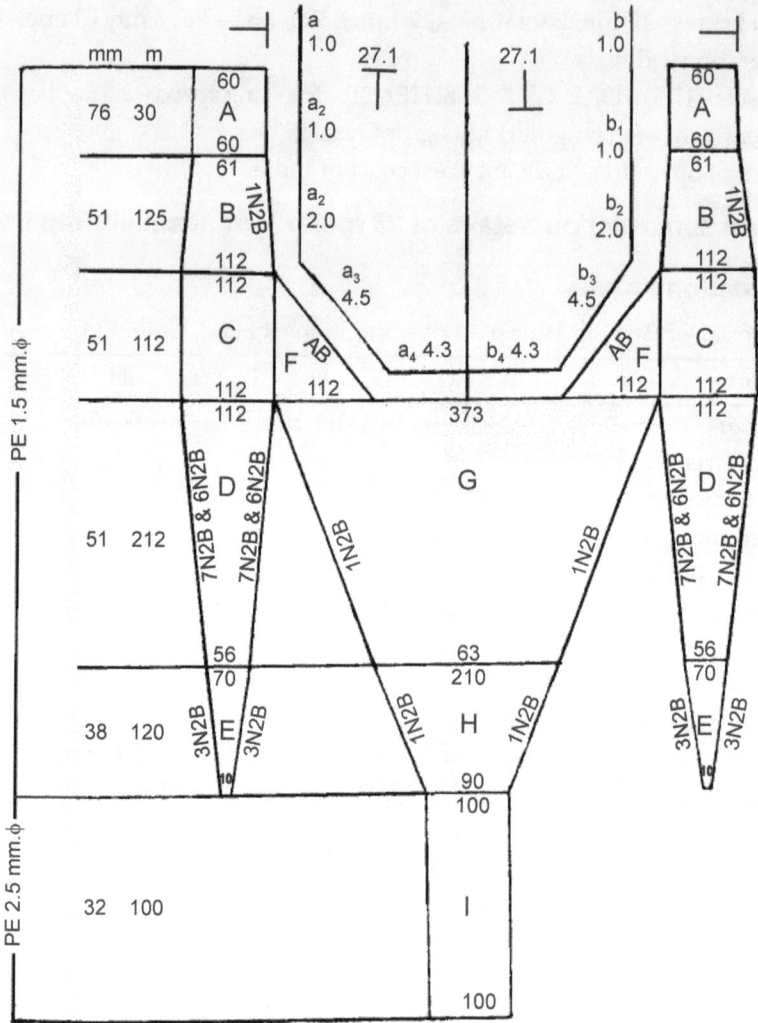

Fig. 19: Drawings of 28 meter four seam shrimp trawl net

Data and construction details of 43.6 meter four seam shrimp trawl net

Construction details

Material – PP or PE; Twine color – Blue

Type of knot – Single sheet bend, Twine diameter – 2mm, Breaking strength – 36 kg

Webbing	A	B	C	D	E	F	G	H	I
Stretched mesh (mm)	<---------------------50-------------------->							42	38
Upper edge (m)	56	100	100	1	482	1	100	120	135

Lower edge (m)	100	100	26	136	100	90	100	120	135
Depth (m)	340	130	294	136	294	136	226	150	195
Baiting rate (inner)	1:10		1:4	1:1		1:1		nil	nil
Baiting rate (outer)		nil			1:2		1:2		

Take up A/B=1/1, B/C = 1/1 D/E=1/1, E/H=5/6, F/G=1/1, G/H=5/5,H/I-8/9

Hanging a1/A= 0.71, b1/A=0.71, b2/B=0.76,a2/D=0.91,a3/E=0.65, b3/F=0.91 b4/G=0.65

Lines and ropes	a	a1	a2	a3	b	b1	b2	b3	b4
Material	Galvanized flexible steel wire rope								
Colnstruction	6x19 with inner fiber core								
Diameter (mm)	<----------- 12------->				<------------ 16--------------->				
Length (m)	1.5	12	6.2	7.2	1.5	12	2.6	6.2	7.2

Floats and sinkers		Floats		Sinkers	
		200		9	
Material		Aluminum &Plastic		GI chain	
Number		18-19			
Shape		Round			
Weight in air (kg)				70-72	

Head rope–43.6 m, Foot rope – 48.8 m, Bolch line –PP or PE rope, 6 mm diameter

Operational details

Vessels – 32.28 mOAL.GRT – 182.6, HP – 578, Trawling speed – 3 knots

Crew – 16, Otter board – Rectangular, 375 kg, Place of operation – Upper east coast of India.

Otter board

Among the trawl gear accessories otter boards are of prime importance and had an evolution parallel to that of net itself. The experiments started in 1860's for an advanced device to maintain the spread of net, required several years of experience for a general pattern to come out. The boards in its present form appeared first in Ireland in 1885, which again took many years of modification for satisfactory' performance. It was in 1892, the first successful otter board trial was made in the English North Sea coast. By 1895, France, Germany and Holland tries these boards and their successes made them popular in all the world fisheries.

A conventional flat otter board is rectangular in shape and made of wood and steel. The length is approximately twice the height and the surface area and weight is suitd to horse power of the vessel. Planks of wood are fitted into a frame made from steel and braced back and front with steel bars. A heavy steel keel is welded to the bottom edge and the lower sides are protected from damage by steel side plates. The triangular towing brackets of which one is slightly smaller than other, are hinged on the front of the board. A pair of iron rings are bolted on the back in a vertical line to which the backstrops are attached. It has become a general practice to use chains instead of one or both the brackets, which allow some adjustment to the rigging of the board. These brackets and backstrops are provided in such a way that when drawn through water the board, by its kite-like connection to the trawl warp, diverts at an angle most suitable for maintenance of spread of the trawl.

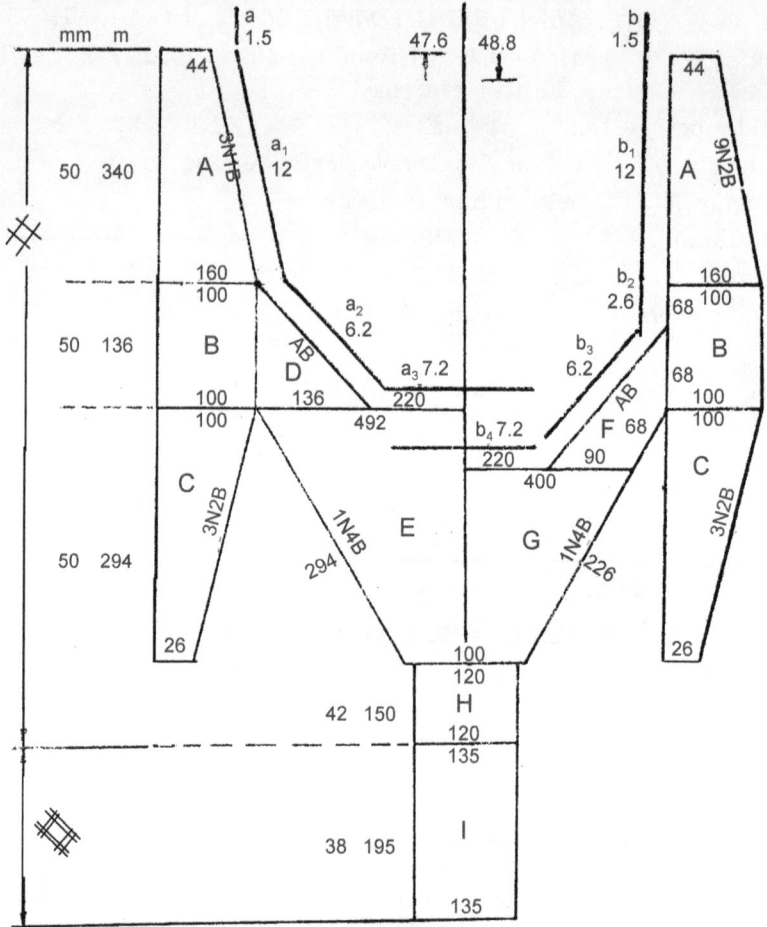

Fig. 20: Drawings of 43.6 meter four seam shrimp trawl net

In the beginning the boards were attached direct to the net, but since the introduction and successful use of Vigneron Dhal patent with its extended spreading wire bridles, the old type is seldom used now. Figure 26 illustrates some of the rigging patterns commonly adopted.

Over the years a number of suggestions have been made for improving the trawl boards and changes have been made in overall size, shape and general pattern and attachment to the warp.The basic idea was to shape the board in such a way that its shearing power become as great as possible, but that the resistance to towing is as little as possible. In recent years, diverse design in shape have been successfully used that include oval, concave, "V" shaped and the cambered type of which oval pattern had been extensively tried (Figs 21-25).

Data and construction details Of flat otterboard-120 kg used in combination with 28 m. shrimp trawl.

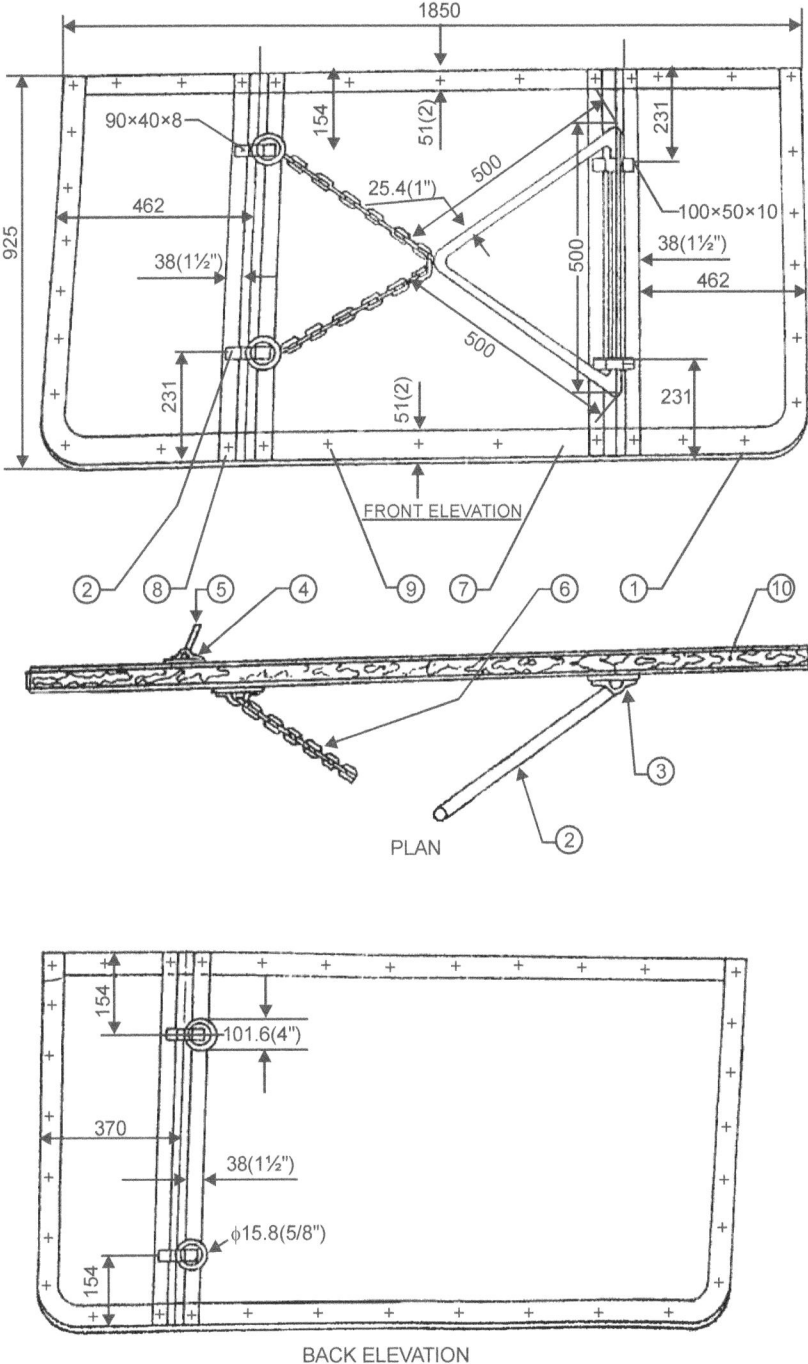

FRONT ELEVATION

PLAN

BACK ELEVATION

Fig. 21 (a) and (b): Rectangular flat otter board-120 kg used in 28 m shrimp trawl.

Area of the board – 1.7 sq. m

Material

Description	No. required	Serial no. in the Fig.
Planks 38 (one and half inch thick) wood		10
Bolts, 9.5 dia. (3/8"), M.S.		9
Middle flats 9.5 (3/8") thick M.S.	3	8
Border flats 9.5 (3/8") thick M.S.		7
G.I. chain link (5/8" dia.) M.S.	1	
Rings M.S.	4	5
Clamp (Backstrop & triangle)M.S.	4	4
Clamps (Triangles) M.S.	2	3
Triangle 25.4 (1") dia. Bar M.S.	1	2
Keel, 12.7 (1/2") thick, M.S.	1	1

Data and construction details of oval flat otter board-50-55 kg for 42 to 56 HP trawlers

Area – 0.7 sq. m

Material	Description	No. required	Serial no. in the Fig.
Wood	Stern board	1	12
Wood	Fore board	1	11
M.S.	Stringers	2	10
M.S.	Triangle clamp	4	9
M.S.	Backstrop clamp	2	8
M.S.	Rings	2	7
M.S.	Big triangle	1	6
M..S.	Small triangle	1	5
M.S.	Keel	1	4
M.S.	Keel sheet	2	3
M.S.	Plate	2	2
M.S.	Plate	1	1

Reflection for right board
Weight – 50 - 55kg
Area 0.7 m²
Propulsion engine power – 42 - 56 HP.
All dimensions are in mm.
(Not to scale).

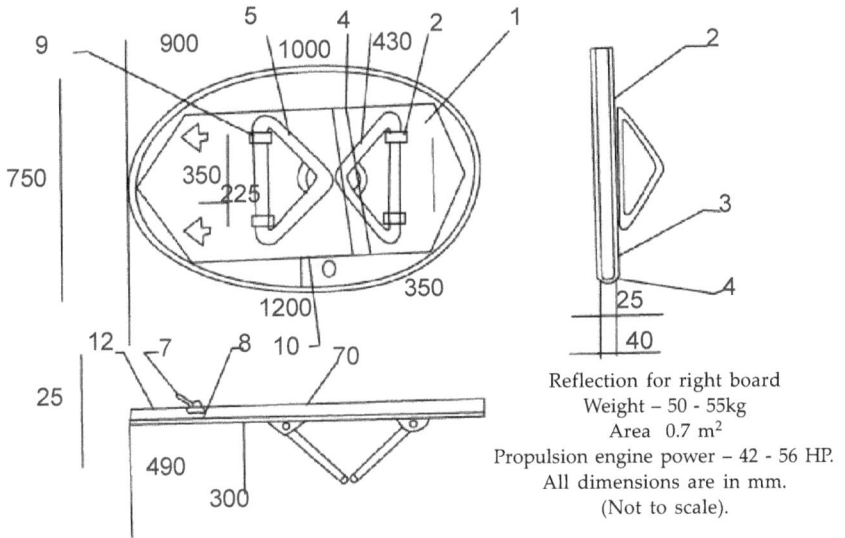

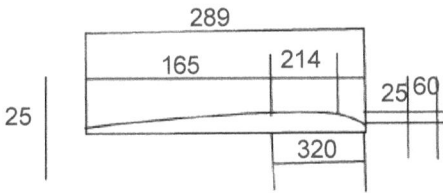

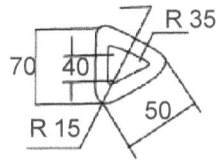

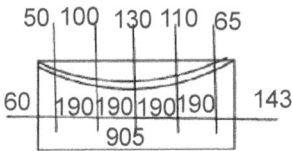

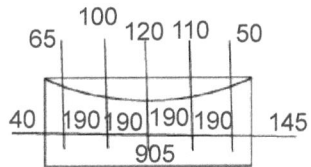

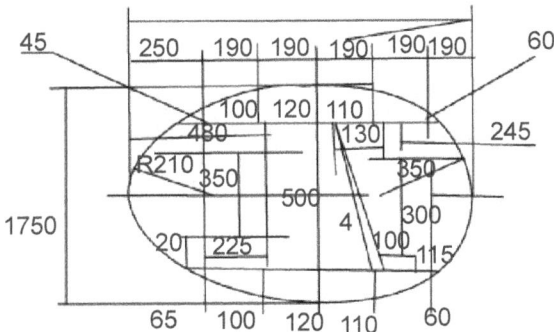

Fig. 22: Oval flat otter board- 50-55 kg for 42-56 HP trawlers

Data and construction details of Oval flat otter board 150 kg for 153-200 HP Trawlers

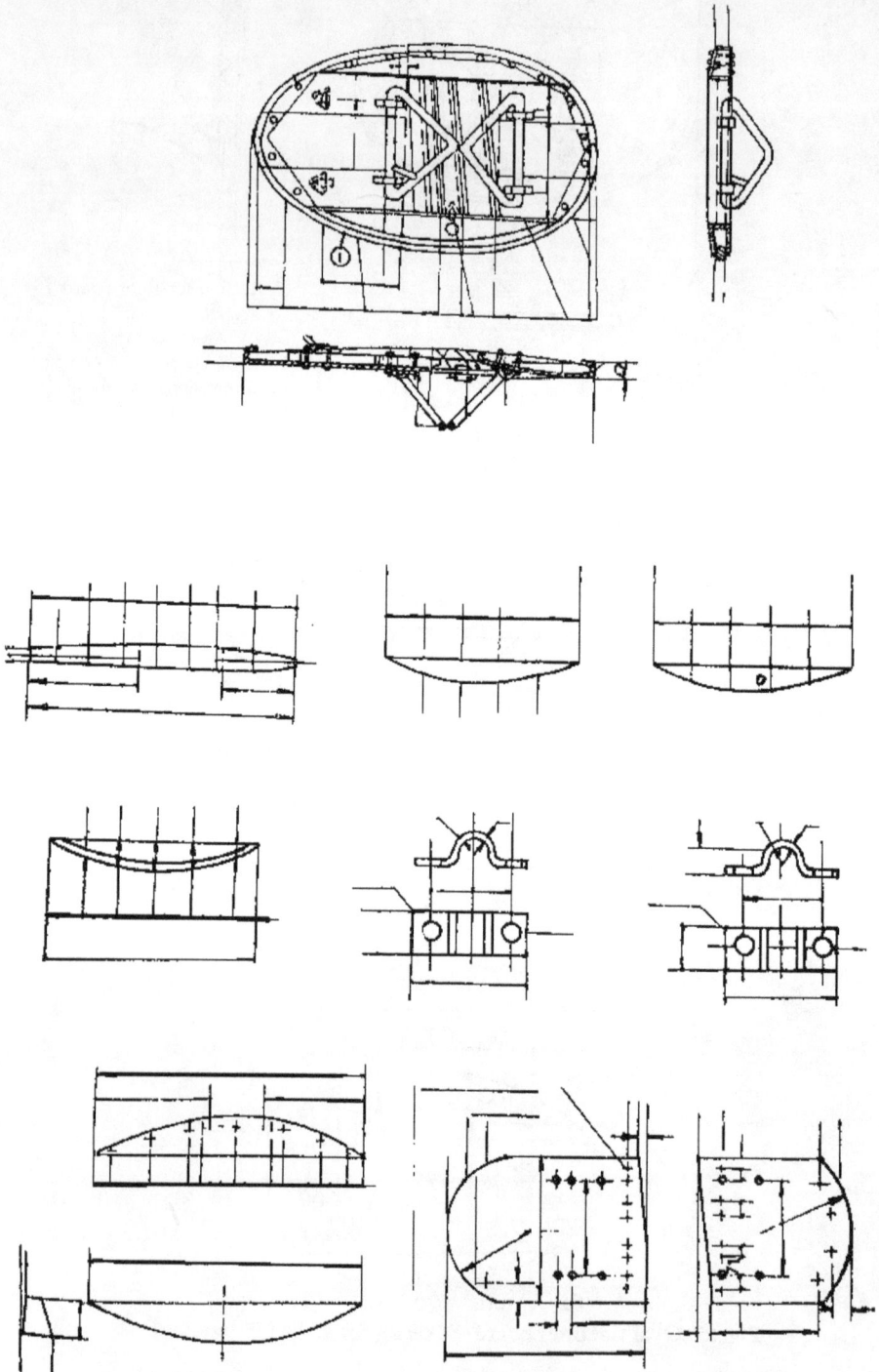

Fig. 23: Drawings of oval flat otter board – 150 kg for 153 - 200 HP trawlers

Area – 1.6 sq. m

Material	Description	No. required
Concrete	Concrete	
M.S.	Washer	4
M.S.	Bolts	12
M.S.	Washer	2
M.S.	Rivets	31
Wood	Upper board	1
Wood	Stern board	1
Wood	Fore board	1
M.S.	Loop	1
M.S.	Backstrop clamp	2
M.S.	Backstrop and lifting ring	3
M.S.	Triangle clamp	4
M.S.	Small triangle	1
M.S.	Big triangle	1
M.S.	Wedge	3
M.S.	Flat bar	2
M.S.	Plate	1
M.S.	Plate	1
M.S.	Plate	1
M.S.	Stern plate	2
M.S.	Fore plate	2
M.S.	Flat bar stern	2
M.S.	Flar bar	1
M.S.	Upper sheet	1
M.S.	Stern sheet	1
M.S.	Fore sheet	1
M.S.	Back keel sheet	1
M.S.	Front keel sheet	1
M.S.	Lower stringer	1
M.S.	Upper stringer	1
M.S.	Keel	1

Fig. 23 Drawing of oval flat otter board – 150 kg for 153–200 HP trawlers

Data and construction details of oval *flat* otter board, 180–190 kg for 200 HP trawlers

Area – 1.6 square meter

Material	Description	No. required	Sr. No. in the Fig.
M.S.	Triangle clamps	4	9
M.S.	Backstrop clamps	2	
M.S.	Backstrop ring	2	
M.S.	Small triangle	1	6
M.S.	Large triangle	1	5

M.S.	Keel	1	4
M.S.	Sheet	4	3
M.S.	Stringer	2	2
M.S.	Stringer and foresheet	1	

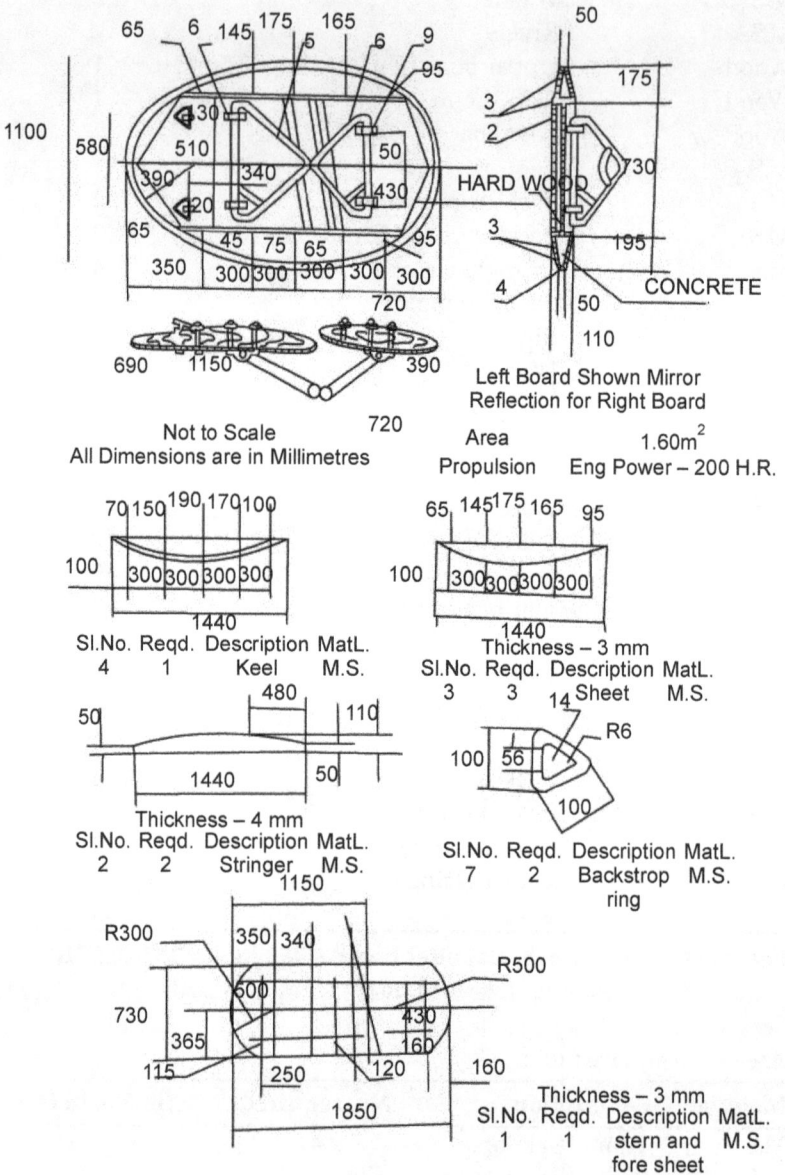

Not to Scale
All Dimensions are in Millimetres

Left Board Shown Mirror
Reflection for Right Board

Area   1.60m^2
Propulsion   Eng Power – 200 H.R.

Sl.No. Reqd. Description MatL.
  4   1   Keel   M.S.

Thickness – 3 mm
Sl.No. Reqd. Description MatL.
  3   3   Sheet   M.S.

Thickness – 4 mm
Sl.No. Reqd. Description MatL.
  2   2   Stringer   M.S.

Sl.No. Reqd. Description MatL.
  7   2   Backstrop M.S.
                  ring

Thickness – 3 mm
Sl.No. Reqd. Description MatL.
  1   1   stern and M.S.
                  fore sheet

Fig. 24: Drawings of oval flat otter board, 180-190 kg for 200 HP trawlers

Data and construction details of oval flat otter board, 250 kg for 240-300 HP trawlers

Area – 2.0 sqare meter

Material	Description	No. required	Sl.no in the Fig.
M.S.	Rod	1	33
Concrete	Concrete		32
M.S.	Washer	4	31
M.S.	Bolt	12	30
M.S.	Washer	2	29
M.S.	Rivets	31	28
Wood	Wood plank	1	27
Wood	Pattern of stern board	4	26
Wood	Pattern of fore board	4	25
M.S.	Loop	1	24
M.S.	Backstrop clamps	2	23
M.S.	Backstrop anf liftingring	3	22
M.S.	Triangle clamps	4	21
M.S.	Limiter	2	20
M.S.	Small triangle	1	19
M.S.	Big triangle	1	18
M.S.	Wedge	3	17
Wood	Plank	2	16
Wood	Plank of upper plating	1	15
Wood	Plank for forged head	1	14
Wood	Plank of upper plating	1	13
Wood	Plank of forged stern	2	12
Wood	Plank	2	11
M.S.	Stern plating	2	10
M.S.	Plating	1	9
M.S.	Upper sheet	1	8
M.S.	Stern sheet	1	7
M.S.	Fore sheet	1	6
M.S.	Back keel sheet	1	5
M.S.	Fore keel sheet	1	4
M.S.	Upper stringer	1	3
M.S.	Lower stringer	1	2
M.S.	Keel	1	1

While assembling otter board special attention are required to be paid to the center of triangles in pair state. Center must be situated, from head of board 850 mm.

The state to be fixed near clamps after fixing. Removing of triangles are prohibited. But removing them down is permitted within the limit of 10 mm.

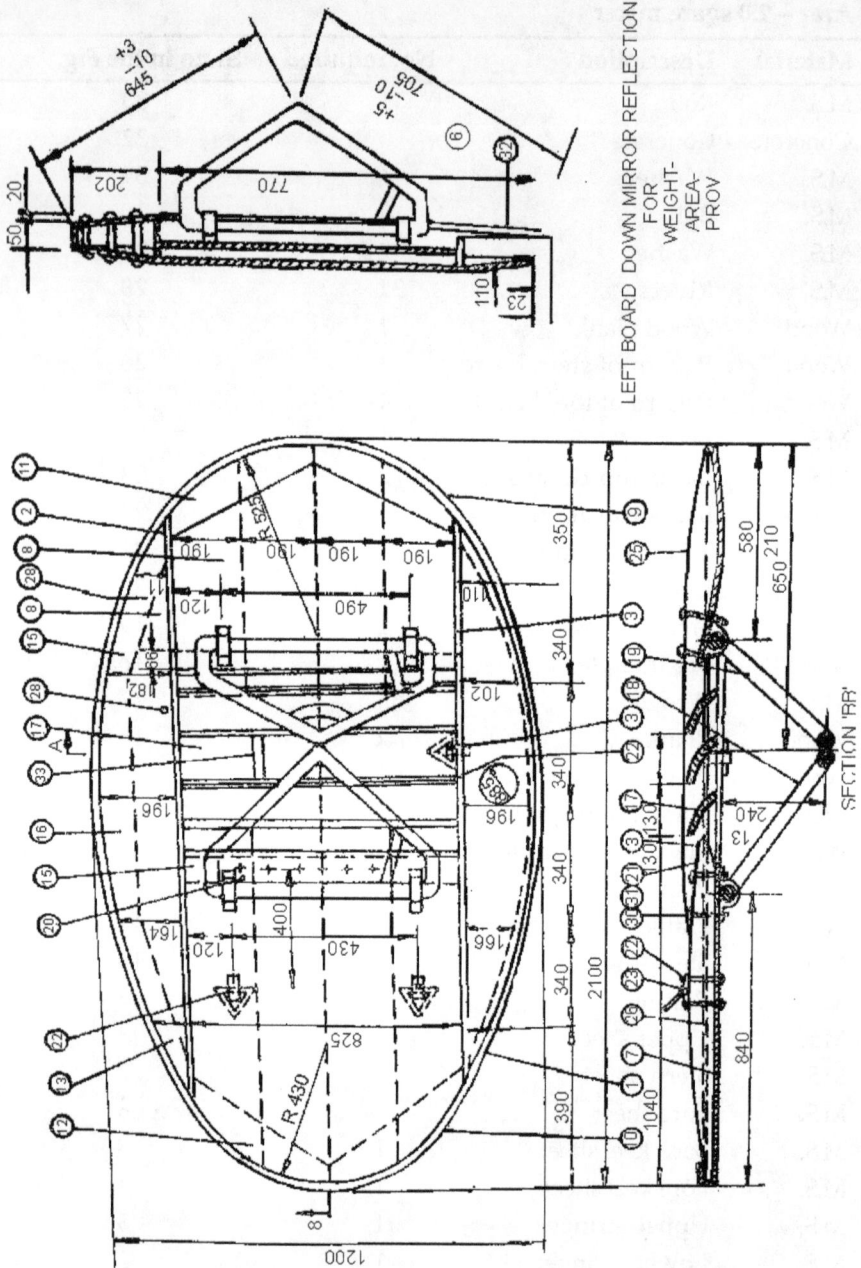

Fig. 25: Drawings of oval flat otter board 250 kg for 240-300 HP trawlers

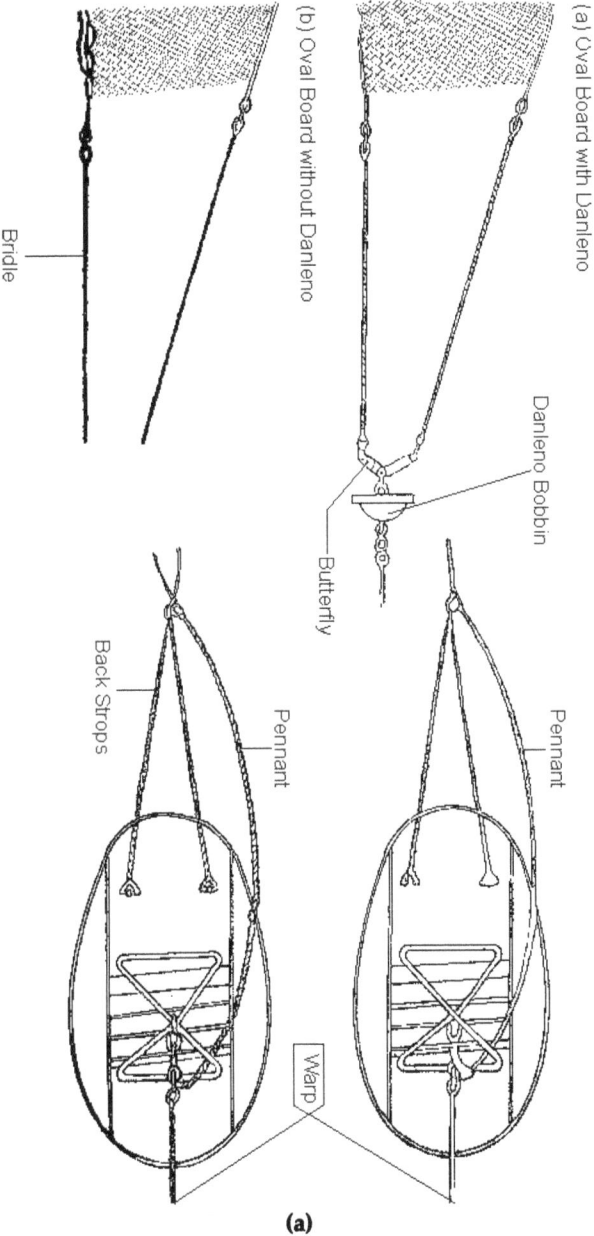

(a) Oval Board with Danleno

(b) Oval Board without Danleno

Bridle

Danleno Bobbin

Butterfly

Back Strops

Pennant

Pennant

Warp

(a)

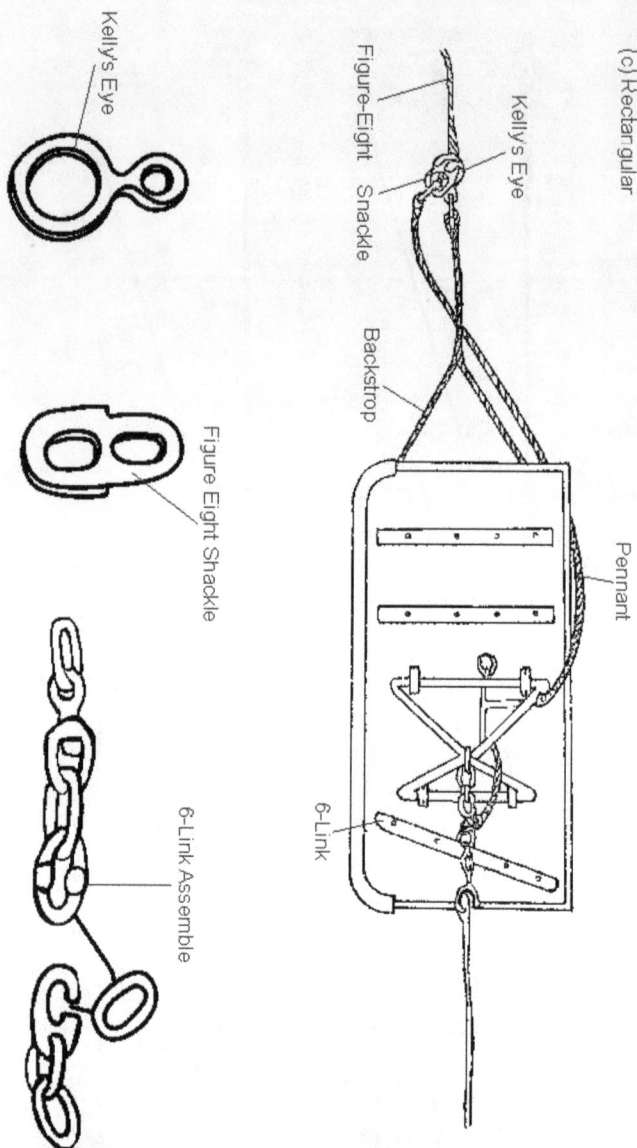

Fig. 26: Some rigging patterns; (a) Oval board with danleno, (b) Oval board without Danleno, (c) Rectangular otter board.

The oval otter boards have many advantages over the conventional type. They have less ground resistance, are less liable to foul on underwater obstruction. The slots, which may vary from one to four produce a smooth flow of water reducing resistance and turbulence. Moreover, it is claimed that they do give a greater spreading power for similar frontal area estimated for single slotted oval otter board at approximately 19% (Garner, 1967).

Otter board size and weight:

There exists no definite principle about the relation between the size or propulsion power of the trawler and the size of the otter boards. H. Miyamoto, an FAO Gear Technologist collected information from Indian trawlers and a few Japanese vessels and found that the area of the board is proportionate to the towing power of the vessel. The relation between these two factors has been expressed by him as;

"S = 0.105 P + 4, where S" is the area of the board and P, the horse power of the engine (Miyamoto, 1959).

A graphical representation is projected in Fig. 27 giving guidelines in respect of area, weight of otter boards for use in vessels of 50 to 500 HP (Garner, 1967). Variations may be necessary based on individual consideration depending on the area of operation and trawler specifications.

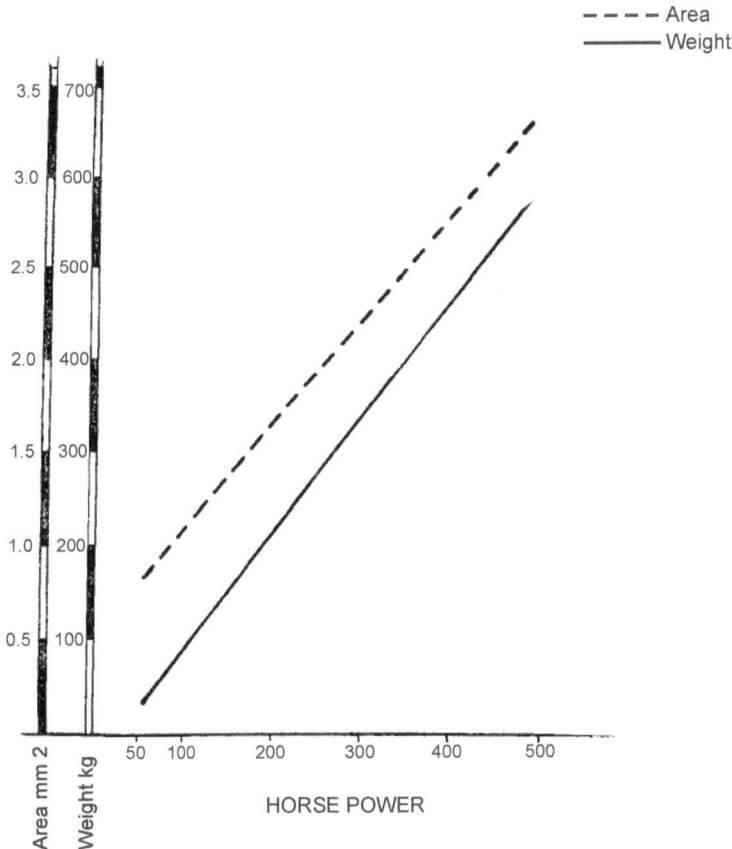

Fig. 27: Estimated engine power ratio to area and weight of otter boards for vessels of 50 to 500 horse power.

Otter trawl accessories

Floats: Floats are essential components of a fishing gear and these obiviously are to help the gear to retain the required position under water maintaining the head lines always upward. The main characteristics of float material are buoyancy, retention of buoyancy, due to absorption of water, capacity to withstand different water pressure and resistance to rotting. Spherical floats of plastic material (PVC) and metallic (Aluminium) floats of different sizes are now increasingly being used instead of cork, thermocole and glass floats

Sinkers: Where floats achieve the lifting of head line, sinkers are used for stretching the ground rope down so as to keep the trawl mouth open and the lower lip to touch the sea bed. Materials most commonly used are iron or pig iron in the form of chain and lead moulded in appropriate shape. The quantum of weight required for different trawls of varying dimensions are indicated in the relevant data sheets But variations of minor degrees may become necessary depending on the difference of materials used, bottom conditions and depth of operation.

Pennants: A pennant or lazy line as it is often referred forms the connection between the figure eight-shackle and connecting link of the warp. It is of galvanized steel wire of same thickness as the bridles and warps. The pennants, one with each otter board are used while the trawl net is being hauled up to take the strain of each sweep line and of the trawl, after the otter boards have been detached. There is no strain on these lines when the trawl is being towed.

Bridles: Bridles or sweep lines are the connecting wires linking the otter board and the leg of the net. In early trawl practices, this part was not in use and the boards were attached direct to the net. But a trawl equipped with bridles is estimated to have a greater spread than the former type. Length of bridles may vary from 10 to 20 meters and the diameter may be same as that of the warps.

Trawl warps: Trawl warps or the towing gear connect the functional gear to the trawler while in operation. They are wire ropes of identical length, two in numbers, fully wound on the winch drum in idle position and are provided with a connecting link at the free end. While shooting, the warp is connected to the otter board by means of "G" link assembly. The diameter of the different wires used in trawling may vary from 11 mm to 16 mm depending on the size of vessel, its towing power and intensity of external forces, like wave action, velocity of current etc. The rope must be flexible galvanized steel, preferably of 6x19 construction with lubricated inner fiber core (6 strandsx!9 wire twisted around fiber core).

Length of warp: The length of warp released while towing the gear is a very important factor principally because the otter boards are to be maintained in the upright position. Theoretically this is achieved at a depth/warp length ratio of 1:3 in 100 fathoms of water. Conditions vary of course, depending on tidal force, any other element, such as, sea bottom etc. But in shallow waters it is often necessary to tow the gear with more than the 1:3 warp ratio and the allowance may extend to 5 times the depth. The equation, $F = (3 + 25/D) D$, where F is the warp length and D, the depth of water, represents an approximate relation between the warp length and water depth (Miyamoto, 1957). For controlling the length of warps as they are paid out from the winch drums and for indicating, some phases of manoeuvring while shooting or hauling warp "markers" are provided (by drawing cotton or synthetic rope of 4-6 mm diameter through the strands of the rope) at every 25 meter intervals.

Additional information: A clear rule has not been evolved so far on the relation between the size of the vessel or its engine power and the size of the net to be used. Miyamoto, in course of his investigations, principally among Indian trawlers found that the size of net used in practice can be related approximately, to the power of engine in the following manner;

$$L = \sqrt{43.6\,P + 660}$$

Where L is the length of the head line in feet and P is the horse power of the engine (Miyamoto, 1959). This project's experience is in conformity with the above generalization.

Bobbins: Rollers or bobbins made of wood, iron, rubber and other hardware materials are designed to roll over the sea bed. Fastened to the ground rope of a trawl they safe guard the lower lip of the net, when fishing over rocks and corals and rough uneven grounds. They may be of various shapes, mainly round, disc or spherical.

Tickler chain: Tickler chain running from one end of the foot rope to the other and slightly smaller in length is some times used to stir up the sea bed in front of the trawl mouth, the disturbing action of which brings up the burrowing and sand and mud dwelling species. It is important in sole and shrimp fisheries

Knotless webbing: First introduced in Japan in 1922, knotless webbings are becoming very popular in many world fisheries and its commercial production started in India in 1964, which is progressively being used in different kinds of net fabrication. Knotless webbing is superior to the conventional type on account of the following reasons;

(1) As there is no knots, less twine is used to make the meshes which considerably reduce the bulk and weight of net.

(2) As the fibers undergo practically no sharp bending, there is no reduction in strength and hence a comparatively lighter net can be used.

(3) Knotless webbing causes less friction in water and hence increase the towing speed.

(4) Due to less friction handling is easy which saves time and labor.

(5) The mesh size is maintained constant because there is no tightening of knots.

Preservation of nets: If natural fibers are used in fabrication of the fishing gear, the net must be periodically treated with preservatives against rot and loss of strength. The action of preservatives are such that while some kill the bacteria, the others make it physically impossible for the bacteria to attack the fiber. There is a wide range of preservatives in use, such as, copper sulfate, copper naphthenate, linseed oil, cutch, coal tar etc of which the first two are recommended for trawl nets. The method adopted for cutch treatment is given below;

For cotton, the nets are first boiled for one or two hours, washed in fresh water and dried. They are then boiled with 3-4% solution of cutch in fresh water or sea water for 2 hours, then dried in the sun or in shade. Upon drying the net is soaked once again in the same solution at room temperature for 12 hours.

While dyeing manila nets, the same method can be employed, but heating should be avoided.

Impregnation with cutch and consequtive sun drying builds up a thin film of tannic acid surrounding the fibers, which prevents the infectious bacteria from entering in. But for a net in continuous use, this acid film will remain only for few weeks and so frequent redyeing will be required. To avoid this, a method has been evolved by application of fixatives, such as, sodium dichromate or potassium dichromate, which can double the life of the cutch treated materials. The process involves soaking of the net already treated with cutch as above, in 1% solution of Potassium or Sodium dichromate at room temperature for one to two hours and then drying. The chemicals may be dissolved by heating, but the dyeing bath must not be heated (Takayama and Shemozaki, 1959).

2 | General Principles of the Design of the Trawl Gear

Trawl net

Trawl is a bag shaped gear towed through water, the mouth of which is kept open by frame, beam, otter doors, kites or floats and sinkers. Some times opening is effected by dragging with two vessels. Trawling, as a new fishing method has recently been introduced in India and trawlers of various sizes are in operation in different parts of the country.

Classification

1. According to the device used for opening the net mouth
   (a) Beam trawl,   (b) Otter trawl and   (c) Bull trawl
2. According to the depth of operation of the net
   (a) Bottom trawl;   (b) Mid-water trawl, and   (c) Surface trawl.
3. According to the method of construction
   Four seam   Two seam
   (a) overhang   (a) overhang
   (b) non-overhang

Parts of the trawl net

1. Towing warp – Connected with otter boards to the winch drum
2. Otter boards – Rectangular or oval boards used for horizontal opening of net mouth
3. Sweep line – Connecting otter board to bridle.
4. Bridle, Leg – Bifurkated sweep line connected to the danleno with two legs.
5. Danleno – Bifurkated sweep line from danleno is connected with upper and lower Part of the wing
6. Wing – Large meshed long trapezium webbings used to scare and drive the fish Into the body of the trawl.
7. Square – The broad portion of the net proper attached to the foot rope for entry of Fish.
8. Belly – Adjacent to head rope the broad portion of the net to receive fish.
9. Flapper – Behind the belly at the beginning of the throat, non-returnable device.
10. Cod end – The receptacle to receive the fish.
11. Head rope & floats – The head rope and floats maintain the positive buoyancy and vertical opening of trawl net.

12. Foot rope & sinkers - Foot rope with sinkers helps to create vertical opening and laying the net during trawling operation

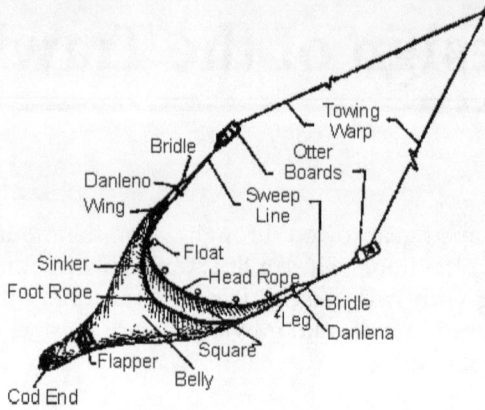

Fig. 28: A typical trawl net and its important parts

Design consideration

The trawl net has to be designed so as to match the resistance of the net with the pull of the boat for effective utilization of the installed engine power. Although the designs of the trawl vary, the different parts of a trawl are proportionate to its size. Miyamoto (1959) after analyzing these relationship has evolved empirically the following formulae for the design of trawl net.

1. The size of the net $H = \sqrt{43.6\, P + 660}$, where, H = Head rope length in feet and P = The BHP of the engine.

2. The length of head rope will be distributed as; H/5 for the Bosom and 2H/5 for the Jibs.

3. The stretched length "L" of the webbings at the upper edge of the belly is, L = H/5 divided by (1-S) X 3, where S = coefficient of hanging.

4 The dimensions of the other parts of the trawl net in relation to "L" are given in The diagram

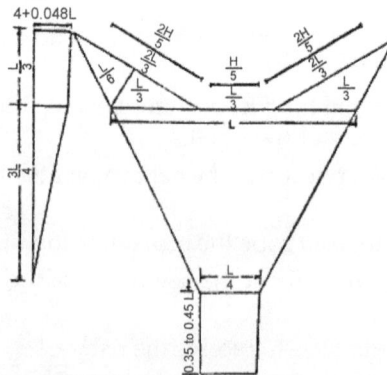

Fig. 29: The distribution of the length of head rope and the dimensions of the various parts.

The mesh size and the twine size required for the webbing of the trawl are determined according to the type of fishery and local conditions. While mounting the net to the rope, the length required for the various parts are calculated according to the hanging effected at the bosom. Generally a hanging coefficient between 0.4 and 0.5 is used. As the horizontal and vertical spreads of the webbing are proportional to the hanging coefficient, some of the corresponding relationships are shown below.

Hanging coefficient	Horizontal spread	Vertical spread
0.38	0,62	0.78
0.40	0.60	0.80
0.42	0.58	0.81
0.43	0.57	0.82
0.45	0.55	0.83
0.47	0.53	0.84
0.48	0.52	0.85
0.50	0.50	0.86
0.55	0.45	0.89
0.60	0.40	0.91

The length of the rope required for the bosom will correspond to the ratio of horizontal spread of the stretched length of the meshes there. But the jib being a triangular piece, the length of the hypotenuse is the required length and can be calculated from the length of the other two sides, which in turn, depend on the corresponding horizontal and vertical spreads of the jib webbing.

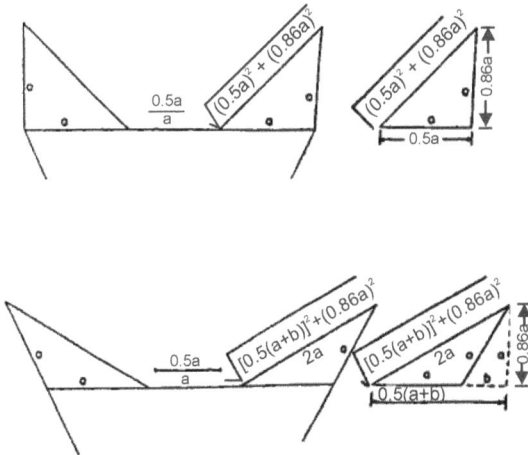

Fig. 30: The length of rope at bosom and jibs at hanging coefficient 0.5

The wing generally consists of straight webbing. So the length of rope will be proportionate to the vertical spread of meshes. In case of tapered wings, the length of the resultant hypotenuse has to be calculated.

Construction

The aim is to construct, (1) a suitably long conical bag which enables water flow evenly and (2) to effect a smooth catenary curve at the mouth.

For this purpose, the webbings required for the trawl net are prepared in sections as per specifications given for the various parts in the design (Fig. 31). These sections are assembled together.

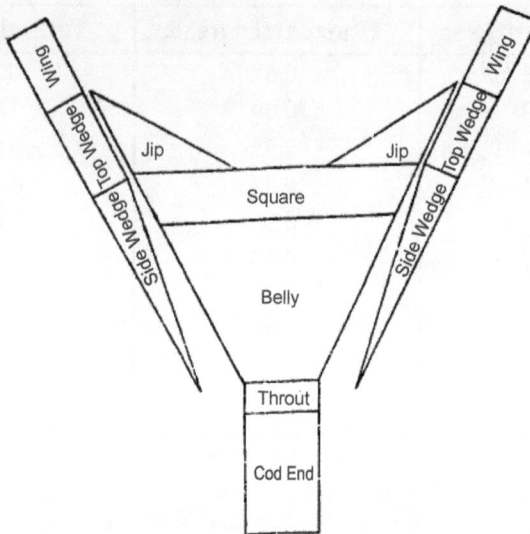

Fig. 31: Various parts of a trawl net design.

According to the construction, the trawl designs can be divided into "Two-seam trawl" (Fig. 32).and "Four-seam trawl" (Fig. 33).

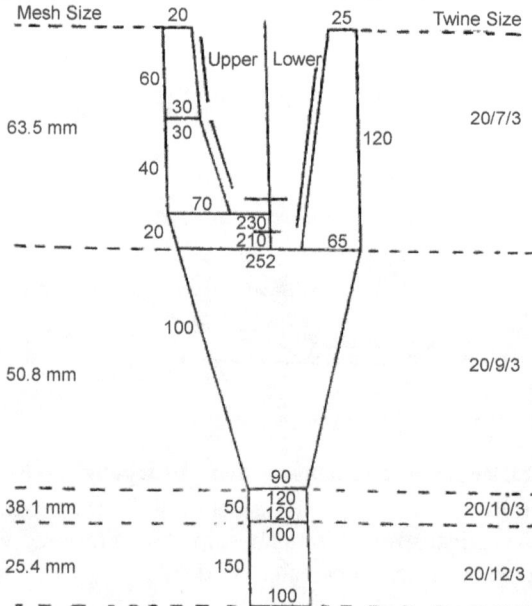

Fig. 32: Design diagram of a two seam trawl.

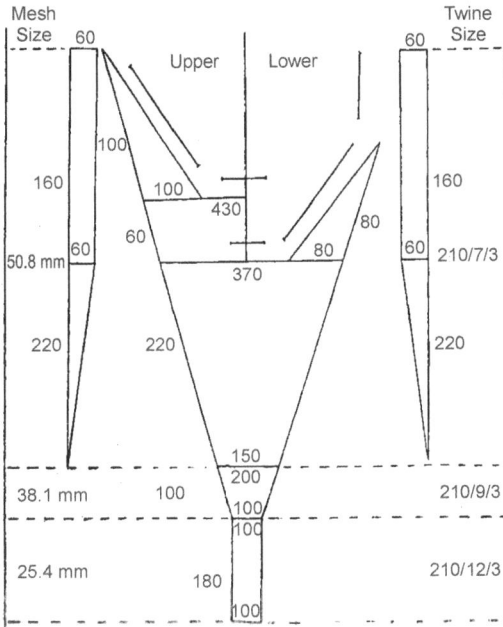

Fig. 33: Design diagram of a four seam trawl.

Larger bottom trawls are of two seam construction. They have only two major parts, namely, one upper and one lower. These nets are normally provided with a overhang for the upper belly.

For the four seam nets, there are four parts, the upper, the lower and the two lateral sides. Nets of this type may be made with or without overhangs. In a non-overhang net both the upper and lower bellies are of identical dimensions. At times, a "square" is separately made and attached to the forward part of a non-overhang trawL to serve the purpose of an overhang. Boats of small and medium sizes generally use four seam nets. Some of the pelagic trawls are also of four seam construction. When the net is intended to catch mainly fish, it is advisable to provide an overhang and for shrimp net an overhang is not essential.

The mouth region of the upper part of the net is mounted to the head rope and of the lower part to the foot rope. Loops of suitable size are made on the rope with hanging twine which is thicker than those with which the webbing has been fabricated and the respective edges of the webbing are then laced to these loops. Another method is to hang the net on the "bolch line" and rig the bolch line along with the webbing to the respective ropes. A third method is to pass a thicker twine of the required length through the edge meshes of the respective parts and this twine with the meshes arranged uniformly is rigged to the relevant ropes. The foot rope is provided with the required number of lead sinkers prior to the rigging of the webbing. The distribution of floats and sinkers to the respective ropes of the trawl needs careful attention. They should be so distributed as to prevent undue sagging of the ropes .For trawls of medium size a satisfactory requirement of weight will be 0.5 to 0.75 pounds of weight per feet length if the foot rope and the total buoyancy of floats required will be between half to two-third of the total weight of the sinkers. Minor adjustments, if any, can be made as per results of trial operations.

After mounting the net, "eyes" are spliced at either end of the rope. The ends of the ropes, also called, "legs" should be of equal length. The length of leg is usually one-fifth of that of the head rope.

The free forward edge of the webbing at the wings is provided with a "wing-tip-rope" entering between the head and foot ropes. A thin strong line is used for the wing-tip-rope, the length of which is limited to the vertical height required for the wing.

Similarly, the free edge of the cod end webbing is further provided with bigger loops made of thicker twine, through which a smooth and strong line, slightly longer than the stretched length of meshes there, is passed and the two ends of the line are joined together by splicing. This is the "cod-end-rope", used for securing as well as for releasing the catches. It is customary to provide a long float line with float to the cod end as a precautionary measure to salvage the net when fouled.

The usual rigs, such as, the "Dan-Ienos", sweep lines or cable, bridles etc. are made ready according to the local conditions or according to the choice of the skipper for the attachment of the net to the otter board.

Cutting rate

The traditional hand braided webbings are being fast replaced by tailored machine made ones in trawl net fabrication. The general principles followed for calculating the cutting rate to tailor a webbing are indicated below.

A mesh has four legs from which two legs or one leg only is cut at a time. When two legs are cut together, it is termed "point cut". There are two types of point cuts; "N-cuts"- when the direction of cutting is at right angles to the general course of netting, and "T-cuts"- when the direction of cutting is parallel to it. "Bar-cuts" or "B-cuts" are made when one leg only is severed at a time from the mesh. In this, the direction of cutting runs parallel to the line of sequential mesh bars.

To obtain the desired taper, the different types of cuts must follow each other at distinct rates in a rhythmical way. The direction of the line of cutting follows approximately the hypotenuse of the resultant right angled triangle formed there and the rate of cutting can be calculated from the number of meshes that constitute the other two sides of that triangle.

Point cuts = P = (L-N); Bar cuts = B = 2 N; where,

"L" is the number of meshes at the longer side of the triangle and

"N" is the number of meshes at the shorter side of the triangle

The rate of cutting is the ratio P/B. If this ratio cannot be simplified completely, the nearest rounded value can be used with discretion.

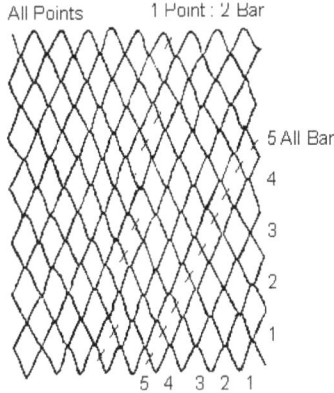

All Points   1 Point : 2 Bar

5 All Bar

4

3

2

1

5 4 3 2 1

Fig. 34: Tailoring of webbing- "Point" and "Bar" cuts

For example, the cutting rate is derived as follows. To cut a taper where 30 meshes are reduced in a length of 60 meshes, the number of point cuts, bar cuts and the rate of cutting are;

P = (L-N) = 60-30 = 30

B = 2N = 2 X 30 = 60

P/B = 30/60 = ½, that is, a total of 30 point cuts and 60 bar cuts are made at the rate of 1P2B.

On the other hand the taper when 30 meshes are reduced at a length of 70 meshes will be;

P = 70 - 30 = 40 ; B = 2 X 30 = 60; P/B = 40/60, that is, 40 point cuts and 60 bar cuts are to be made and the rate will be 1P1B alternated with 1P2B so that the different cuts are distributed uniformly

Trawl nets - design and development:

Granton Trawl Nets or extended version continue to represent a large proportion of the gears currently in use in Britain. Similarly modified Grantons also remain in demand in many parts of the world. It is accordingly appropriate to examine the basics of the Granton Trawl Net pattern, itemizing each section in turn inorder to clarify fundamentals of trawl net design and assembly.

In the following paragraphs therefore, the theory and principles involved in the initial pattern are examined. And later, those for extending existing designs for possible increased productivity.

Otter trawl nets were developed from the beam trawl and designed with seven main pieces.

1. The square: The section of netting fitted between the top body and the two top wings, so that it partially overhang the ground rope. It is really this part of the trawl net which is most important, for it does to a great degree govern the way the rest of the net can be shaped.

Specification includes, 260 meshes of double twine one set up in a 5 and half to 6 inch mesh and two double rows are braided. Continuing with single twine, baitings are made at each side every fourth row decreasing to 200 meshes across.

If a 5 and half inch mesh is required, the finished length would be approximately 28 feet, whereas 6 inch mesh would result in a length of 30.5 feet. The cutting rate on machined netting would be 1 point 2 bars repeat.

Machine piece would be set up with 59 meshes (shuttles) across the loom and 460 rows (230 meshes) would be manufactured. The small taper cut from 30 meshes to 1 mesh would then be reversed and rough joined on to the last row made to produce the finished piece as required 260/200/120 rows after two rows have been hand braided on to the wide edge. The multiples would be repeated for say ten squares, but only the last piece would have a rough join, following alternatively reversed cuts of 260 meshes to 200 meshes. On a hundred shuttle machine it would be possible to plan for one and a half widths (selvedge length) with the first half tapering from 260 meshes to 230 meshes and the second decreasing the rest of the way from 230 to 200, thus a number of shuttles would be used to make 89 meshes with one half mesh lost when cutting at 59 meshes for one and a half square selvedge lengths.

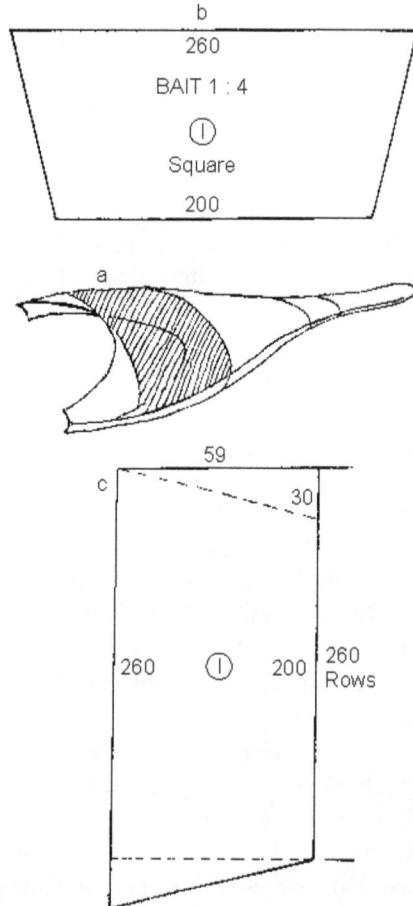

Fig. 35: Diagram of the square.

2. The top wings: Two sections of netting usually shaped diagonally opposite to one another to form the upper mouth of the trawl net. The head line is attached from one top wing end along the diagonal fly meshes across the bosom or center part of the square and along the opposite top wing to the end. The quarters are the two areas where each of the top wing fly mesh edges join in the square. With conventional trawl nets the quarters are usually subjected to heavy loading due to inadequate shaping, and this is particularly true for such a trawl net as is operated from a side trawler when the gear often has to be shot away with uneven strain on the head line.

Specification include 90 meshes are set up in a 5 and half or 6 inch mesh making 15 or 20 in double twine at the inner fly mesh side which forms the double twine at the inner fly mesh side. This form the "double corner" (quarter area) which is usually made first, then the wing is completed to 11 meshes wide, mostly in single twine, following with fly meshes from the double corner. This reduces the width by one mesh for every two rows braided, as the outer selvedge is plain, which means, if there are no baiting in the quarters that the finished length would be 36.5 feet in a 5 and half inch mesh or 40 feet in a 6 inch mesh. However, it is normal practice to make sixteen or so baiting in the fly mesh edge, then the length would be reduced to 29 feet and 31.5 feet respectively. It is also the practice to include in the finished length several double rows at the wing end. And occasionally two double meshes may be braided along the entire fly mesh edge. In each case, the purpose is to gain added strength for the more vulnerable parts of the netting.

Machine piece – Assuming baitings are required in the top wing, they have to be discounted when planning machine manufacture, for no loom will make baitings. The 16 baitings therefore would have to be deducted from the base width of 90

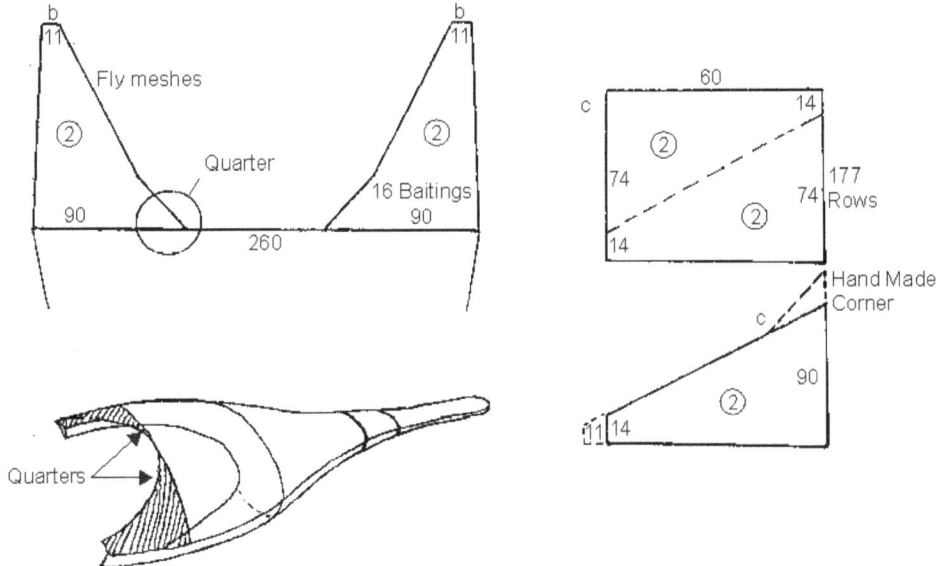

Fig. 36: Diagram of the top wings

meshes leaving 74 meshes. A length of 29 feet or 126 rows in 5 and half inch mesh may then be planned in the width of the machine with 60 shuttles leaving six rows for doubling by hand, meaning a loss from 74 meshes to 14 meshes. Thus the piece for two top wings would be planned 60 meshes wide by 177 rows- one row being required for cutting the diagonal wings

3. Lower wings: Two narrow sections of netting fitted between the lower belly and the top wings to form the lower lip of the trawl net. The ground rope is attached from one end to the other, along the fly mesh edges and across the lower belly bosom meshes. The two lower wings are the parts of netting which are subjected to the most abrasion and consequently they are the sections which have to be continually repaired or replaced when working rough ground.

Specifications – Fifty meshes are set up in a 5 and half inch or 6 inch mesh making 10 in double twine at the inner fly mesh side for the double corner. Fly meshes are made and the outer selvedge is creased every third row, thus in a 5 and half inch mesh to reduce from 50 to 17 meshes, which is the required width, the length would be approximately 45 feet as the loosing rate is one mesh for every six rows made.

In a 6 inch mesh the length would be 49.5 feet, but irrespective, a double row has to be made at this juncture to mark the part where the wing end section begins with 17 meshes maintained throughout by fly meshing and flat baiting (baiting every two rows) to the required length which should be 10% greater than the total stretched length of the square and to wings including several double rows at the wing end. Again double meshes may be made along the entire fly mesh edge as added strengthening.

The "bunt" and its actual specific dimension is frequently mistaken for it may be regarded as the area up to the double round marking the beginning of the wing end, which in the case of the 5 and half inch mesh section would be 45 feet long. However, this can be confusing and in fact the "bunt" is really the netting between the bosom corner up to the end of the bunt bobbins or bunt section of the ground rope. In other words a double row should be made or a colored mark inserted for several meshes at a position approximately 15% greater on the netting than the length of the bunt bobbin wire or similar part of the assembly. For example, in the case of a 20 feet ground rope or foot rope wire being in this position the marker would have to be 23 feet from the start of the lower wing which would leave about 55 slack netting to be spaced along the remainder of the ground rope.

Machined piece- It is not convenient to machine only small numbers of lower wings because of their length and general shape, as unsuitable rough joins have to be made across a good deal of the length of each wing. However, on the 101 shuttle loom 99 shuttles would be set up to make the body of two lower wings with 135 rows produced. The required wing end section can then be hand braided on.

The fly meshes for both top and lower wings which have been produced by the machine have to be added later by hand.

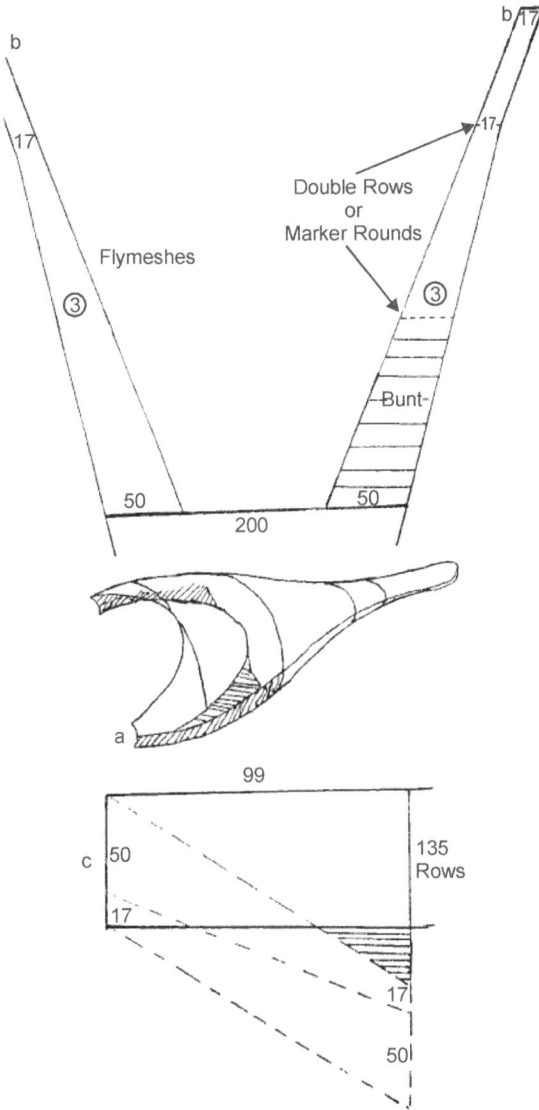

Fig. 37: Diagram of the lower wings

4. The Bellies: The bellies are the channel of the trawl which guide the fish along into the cod end. Two are required for each trawl and the top one used to be referred to as the "baiting", whilst the lower was regarded as the belly of the trawl. Conservation aims and changing minimum mesh legislation have meant that this part of the trawl net has continually had to be changed. This compulsion has not, in fact, always been as bad as it sounds for the increases demanded have led to the trawl net being gradually enlarged. This has happened over a number of years and at the same time the size, range and power of vessels has steadily increased.

Depending on the twines used, the minimum mesh size (inside measurement) is between 110 and 130 mm for the normal deep sea regions, which the heavy British trawlers work .Assuming this that 5 and half inch mesh is worked in the mouth of the trawl net, this should be continued along the entire length of the bellies, whereas a few years ago the belly was graduated into a number of spools of decreasing mesh size. A Granton Trawl, of course, may be used for herring fishing in an area where the legislation is not applicable, when the graduations in the belly may drop in panels to a two and half inch mesh. For practical purposes however, a pair of bellies can be considered to have an overall length of 50 feet with approximately a 5 and half inch mesh throughout.

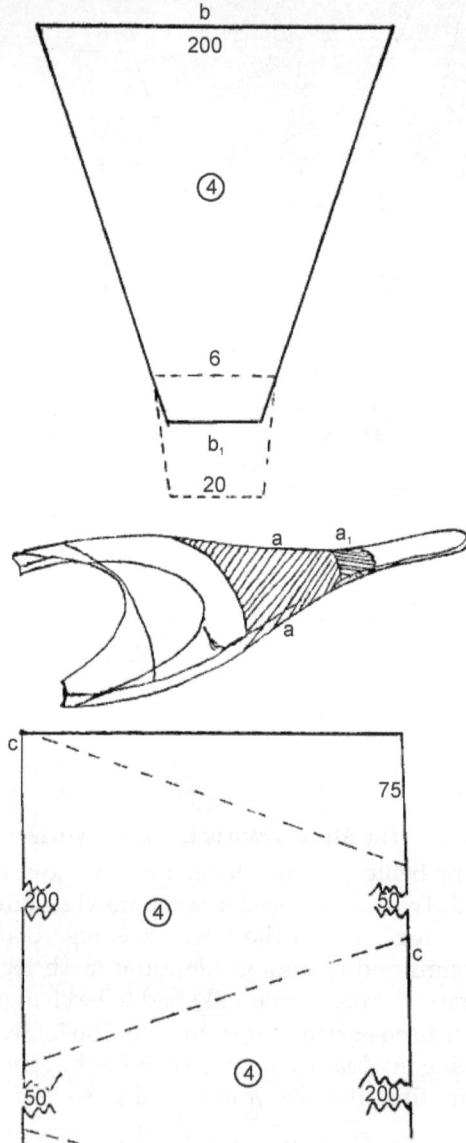

Fig. 38 : Diagram of the Bellies

Specification – 200 meshes are set up with double twine and 2 to 10 double rows, or in certain cases even more, are braided, and continuing with single, baiting are normally put in every third row at each side down to say 50 meshes. The length therefore, each belly would be with a 5 and half inch mesh, 51.5 feet (225 rows, 15 baitings every third row). Although there are no changes of "spools", it is advisable to have double rounds or several colored marker meshes dividing the belly up into sections in order to facilitate repairs.

Machined piece— As is apparent from the "specification" to achieve the taper 1:3, the length of the bellies is 51.5 feet or 225 rows which is 112 and half meshes/ shuttles. This sort of requirement is perhaps the simplest section to plan for manufacture provided the width of the loom is sufficient. Assuming however, that one of the more common 101 width machines were to be used, then it would seem appropriate to plan for the full width of the machine and offer a belly finishing at 100 meshes long on the selvedge with a finished length of almost 46 feet or to braid on at the tail end of the belly to a total length of 225 rows. In this case where 10 double rows are required at the head of the belly then the question would be solved automatically by adding appropriately.

5. Lengthening pieces: Originally, lengthening pieces simply served the purpose which the name applies. They were 10 to 20 feet sections used to extend bellies as required. Now a days, however, this extension may be allowed for in the belly length, or it can be used to advantage in certain other cases.

Specification – Using double netting for strength an extension of about 15 feet might be added to make a section from 66 to 50 meshes decreasing with baitings made every eighth row at each side. Total number of rows 64.

Machined piece – If the length of the belly was to be planned with the normal 1:3 taper ratio to facilitate 101 shuttle loom manufacture so that the length of the belly was only 46 feet with 66 meshes at the narrow end and a greater body trawl net length is required for stern trawling, then a lengthener could be added to advantage.

6. Cod end: Now a days cod end are two rectangular pieces of netting manufactured with heavy double twine. The top edges are joined to the narrow end of bellies, the selvedge are laced together and a cod line or cod end clip is reaved through the lower meshes for securing the section into a bag where the fish are held until released on board the trawler. It is the cod ends which are most affected by mesh legislation and the size of the cod end depends on the size and type of the vessel, the area to be fished and the species to be caught.

Specification – The minimum inside mesh size allowed to be worked in the North Atlantic deep sea fisheries varies between 110 and 130 mm. Also a stern trawler will work on longer cod end than a side trawler. For practical purposes, however, the specification can be regarded as follows.

Set up 50 meshes of double twine and braid 174 rows with α 5 and half inch mesh plain down to 50 meshes.

Machine piece – Provided the machine is of the heavy type of 101 shuttle width, 67 meshes would be set up and 100 rows manufactured for each piece.

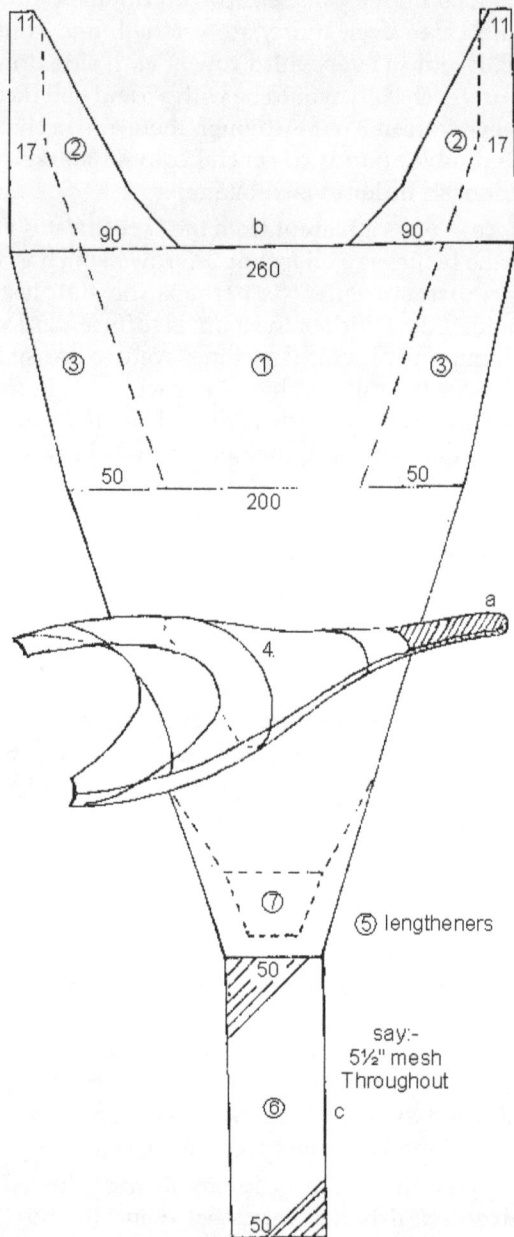

Fig. 39: Diagram of lengthening pieces and cod end

7. Flapper: The flapper is a small single piece of netting tapered and fitted into the belly near the cod end, so that after the bellies have been sealed with twine diagonally towards the selvedges with the top edge of the flapper laced to the top belly it acts as a non-return valve after the fish have entered in the cod end.

Developing conventional otter trawl net

A large proportion of development from beam to otter trawls and onwards have been done empirically and the Granton Trawl net has been, in its basic design, an extremely efficient harvester of the sea. The great advantage with this gear was its degree of standardization to suit particular sea floor environment and conditions in North Atlantic waters which were being fished by average vessels of certain class.

To exemplify, looking back between the mid- 1940's to the early 1960's it would not be exaggerating to say that the Granton Trawl net fitted with specific types of assembly gear and standard doors, with which crews had grown familiar proved extremely satisfactory as worked from the deep sea side trawler in the 1000 to 12000 HP range. However, as trawlers have changed, growing larger power wise and dimensionally, there has been more practical experimentation for an overall increase of the trawl gear parts. This is understandable but not always successful.

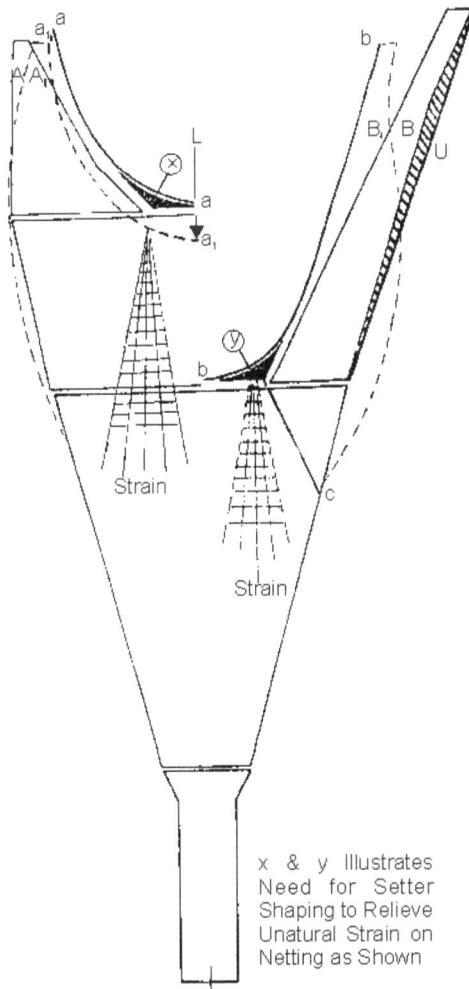

Fig. 40: Dimensional transfiguration of the "Granton" trawl mouth area

To examine Fig. 40, which illustrates the Granton Trawl net successfully fished over the period indicated, it will be noted that the net pattern is shown as it would appear in plan form and then in the shape it might adopt (dotted lines) under operational conditions with the trawl on the netting at positions X and Y. As the lower wing has certain percentage of slackness, the strain at Y will not be so severe as that experienced in the area X. Neverthless the shaping of these corners can be easily improved.

Generally in fact, the shape of this trawl net does not allow for a great deal of manoeuvrability and this is certainly not helped by the narrow strip of netting which makes up the lower wing end. There is no reason why the specification for this part of the netting should not be made to average out to the wing end as shown in the shaded section U. This may not appear severely restricting from the drawing, but when extending the Garaton Trawl net pattern with deeper lower wing ends, it does not become an important consideration..

Similarly a further example of comparing trawl net shape is ahown in Fig. 41. The entire line divides the lower sections A from the upper sections B with the mouth areas shown over C and the body areas shown over D. The shaded part of the drawing illustrates the netting plan form (to scale) of the standard Granton Trawl with cut away lower wings, against the doted lines of the square, top wing and lower wing of the trawl net commonly known as French Trawl, which was introduced into Britain about 57 years ago.

Now considering the French Trawl manufactured with light nylon twine was initially operated from only medium class trawlers, whilst the Granton Trawl was worked from large stern trawlers, it must be obvious that there is a good deal of scope for practical development, particularly in the wings of otter trawl nets. The emphasis of the French Trawl was on the top wings which were fly meshed and baited every other mesh along the entire length to give the mouth of the trawl net a wider gape or improved spread, provided the otter boards and other assembly gear were correctly balanced.

There are hundreds of variable trawl net designs to suit each particular class of vessel, but basically, dealing with the deep sea type of gear, the design usually be traced back to resemble the original Granton Trawl net pattern. This is with the exception of last 56 years when a good deal of original thought has been put into this subject by a few experienced fishing gear technologists.

Accordingly, and assuming that the Granton Trawl is to be extended to suit a stern trawler of 2000 HP, then it might be decided that the head line should be increased to 100/120 feet. The square may be left 260 meshes although the size of mesh may be fractionally increased. The base of the top wing would possible to remain at 90 meshes, which means the bosom lenth or head line would be basically unaltered. The length of the top wings, however, would have to be increased proportionately, and if 11 meshes are to be maintained at the wing ends then the creasing rate would have to be increased, which would automatically mean more nettings in the top mouth area of the trawl. Similarly the lower wings would have to be extended with the proportional values, explained in the previous paragraphs, but the taper would extend to the wing ends, in other words, the shaded area "U" on Fig. 40 would be allowed. In this way trawl gear can be enlarged to suit vessels of particular power. It should be mentioned, however, and this will be noted in later chapters of advanced gears, that there is a danger of over loading the netting when increasing the size of trawl nets. For instance, it has been established that the drag on a conventional Granton trawl net is usually, undr normal conditions, about 50% of the whole load of the gear. It is therefore logical to assume that to increase trawl net size in traditional ways could add fairly heavily to this percentage. Again this is where the increasing minimum permissible mesh size is helping to some extent. This sort of problem is sometimes referred to as "below out" and in addition to improved shaping the answer seems to lie in the direction of increasing the mesh size in the wings and careful gradation down the body netting. Light twines are also advantageous within reason, provided the netting is correctly supported with light strengthening ropes of sufficient numbers.

The question which often arises on trawl gear design relates back to that shown by Fig.40, for why should a small boat tow a larger net than the big trawler etc ? Once again the answer often lies in the size of the twine and the way the netting is loaded with accessory gear, but the real solution is obtainable from more practical research and development under operational conditions.

Size of trawl gear matching to engine horse power (HP).

Most of the modern Japanese trawlers are stern trawlers, which range widely from 300 to 4500 gross tons. An important problem for trawl technologists to day is how to seek the size of trawl fit for the horse power of a main engine. Studies on this problem for the past several years, this chapter introduce computation methods for power to pull trawl and proper size of related gear such trawl winch, trawl net, otter board and warp for trawl technologist's reference.

Power to pull trawl

Power to pull trawl is influenced by the horse power of a main engine, shape of trawling vessel, and structure of propeller including its diameter, its pitch and its developed blade area. It is also influenced by trim of trawling vessel under operation. It is very difficult, therefore, to compute an exact power to pull trawl. Strictly speaking, each trawler has its own power to pull trawl. If the exactness is too much required, finding a rule for computing this power is almost impossible. Therefore, a method for estimating the approximate value of power to pull trawl that has been established by a number of experiments have been given.

Six trawlers selected for studies are from 300 to 3500 gross tons and are built in Japan between 1964 and 1967. The principal dimensions of the trawlers and propellers are given below and the relation between power to pull trawl (EHP) and shaft horse power (BHP) found by experiments is shown in Fig.41 (in the calm sea with the wind at Beaufort 1-3). The power to pull trawl or EHP is horse power necessary for pulling the gear on the condition that hull resistance is not calculated.

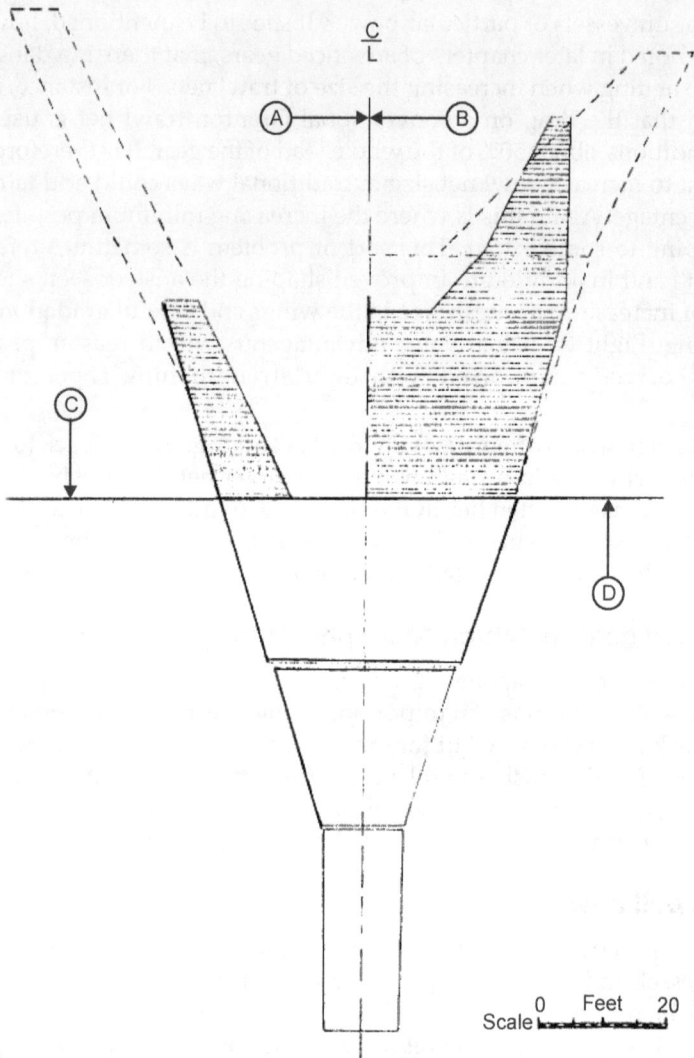

Fig. 41: Comparison of trawl net shape

Table -1: Summary of experimental trawlers

Name of trawler	A	B	C	D	E	F
Length,LOA (m)	39.80	73.79	88.20	88.20	88.20	99.50
Breadth (m)	8,00	12.10	14.00	14.00	4.00	15.50
Depth (m)	3.90	5.75	9.20	9.20	9.20	10.00
Main engine type	Nigata	6 UET	7 UET	7 UET	7 UET	8 UET
Max.continuous BHP	1200	2700	3150	3150	3500	4000
Max.engine speed(rpm)	550	225	225	225	235	240
Gross tonnage	314	1500	2800	2900	3000	3400

Propeller type	CPP	Fixed	Fixed	Fixed	Fixed	Fixed
Propeller dia.(mm)	2400	2950	2950	2950	3040	3150
Propeller pitch (mm)	1440	2050	2210	2220	2100	1970
Propeller	1936	4580	4440	4440	5382	5058
Developed blade area						
No. of blades	3	5	5	5	5	5
Trawl winch capacity	8 ton	15 ton	18.6 ton	18.6 ton	19.7 ton	20 ton
Winding speed m/min	80	70	60	60	66	80

A – No. 3 Heian Maru; B – No. 83 Taiyo Maru; C – No. 81 Taiyo Maru; D – No. 82 Taiyo Maru; E – Zuiyo Maru; F – Shoyo Maru

When the total resistance of gear is R (kg) and towing velocity is V (m/sec), the EHP is RV (kg, m/sec) in horse power. As 1 horse power is equivalent to 75 kg. m/sec, the EHP is calculated as RV/75. The shaft horse power or BHP is called brake horse power, which is obtained by multiplying indicated horse power (IHP) by engine efficiency. The indicated horse power is obtained by measuring the gas pressure inside a cylinder with an indicator. According to Fig. 42, the relation between the EHP and the BHP of each trawler is almost proportional, and accordingly it is expressed as follows:

$$EHP = K(BHP) \qquad \qquad ...(1)$$

When the horse power of main engine is 1200 PS and the diameter of propeller is 2400 mm, the value of K is about 0.18. When the horse power is 2700-3150 PS and the diameter is 2950 mm, it is about 0.22. When they are 3500 PS and 3040 mm respectively, it is about 0.27. When they are 4000 PS and 3150 mm respectively, it is about 0.3. As the horse power of the main engine increases the diameter of propeller increases, and relative value of EHP to BHP becomes greater.

In indicating the horse power of a main engine, the maximum continuous shaft horse power is adopted an Japan. Generally, about 60% of this maximum continuous shaft horse power is used as BHP during trawling. Of course, more than 60% may be used in a rough sea operation, but in computing the EHP it is recommended to take 60% as standard.

As an example, suppose the horse power of main engine is 3150 PS and other conditions, such as diameter of propeller etc. are those given in Table 1, the BHP used during trawling or 60% of 3150 PS is 3150 PS x 0.6 = 1890 PS. According to Fig. 42, the value of K in the case of given conditions is 0.22, so from formula (1);

$$EHP = 1890 \text{ PS} \times 0.22 = 415 \text{ PS.}$$

Thus the power to pull trawl when the horse power of main engine is 3150 PS is calculated at about 415 PS.

Since 1 PS is equivalent to 75 kg.m/sec,

$$415 \text{ PS} = 75 \text{ kg.m/sec} \times 415 = 31.1 \text{ tons m/sec.}$$

The towing velocity required in calm sea is 4.5 knots (2.25 m/sec), so the resistance of the whole gear at this speed is as follows :

31.1 tons.m/sec divided by 2.25 m/sec = 13.8 tons approximately.

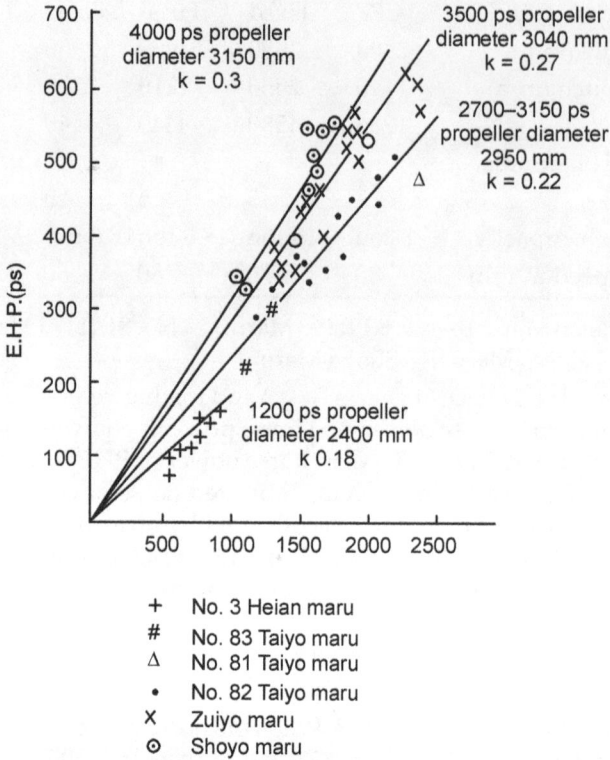

+   No. 3 Heian maru
#   No. 83 Taiyo maru
Δ   No. 81 Taiyo maru
•   No. 82 Taiyo maru
X   Zuiyo maru
⊙   Shoyo maru

**Fig. 42: Relationship between shaft horse power (BHP)
and power to pull trawl (EHP)**

The result of the above calculations show that the trawl gear should be so designed as to have its total resistance of 13.8 tons approximately at a towing

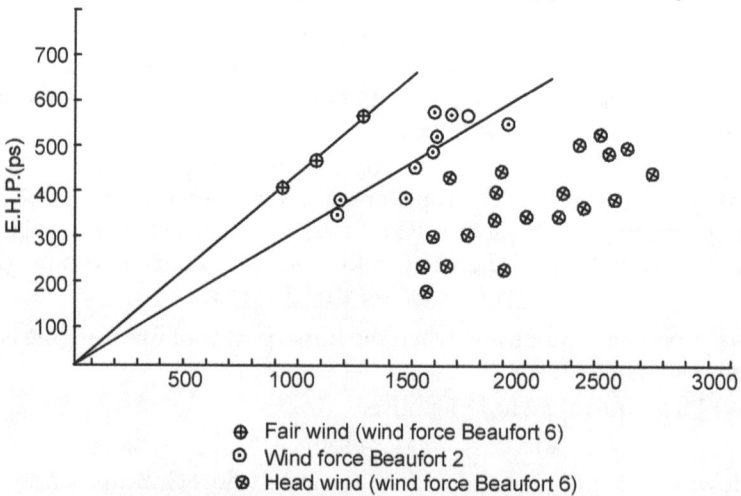

⊕   Fair wind (wind force Beaufort 6)
⊙   Wind force Beaufort 2
⊗   Head wind (wind force Beaufort 6)

**Fig. 43: Relationship between shaft horse power (BHP)
and power to pull trawl (EHP) at rough sea.**

speed of 4.5 knots. This towing velocity 4.5 knots is so determined in order to leave something in reserve for trawling in rough sea where the resistance of the hull increases Fig. 43 shows EHP in rough sea. In Fig.43 it is observed that the EHP in rough waters with the head wind at Beaufort 6, drops down to about one half of the EHP in calm sea. From this fact it can be said that the towing velocity 4.5 knots in calm sea decreases to about 3.5 knots in rough sea with head wind, because power to pull trawl or EHP increases in proportion to the cube of towing velocity, i.e. 3.5 knots cube/ 4.5 knots cube = $^1/2$ approximately.

Trawl winch: It is no exaggeration to say that trawl winch for trawler is as important as the main engine. When hauling operation is started in rough sea, the tension imposed on trawl warps increases abruptly due to scooping action of swell, from the moment otter boards detach from the sea bottom, and the transient maximum tension becomes 3 to 4 times the tension during towing operation. This should be remembered in designing a trawl winch and sufficient allowance should be given to its capacity. If it is driven by electro-motor, a low speed motor with great torque, that is, a motor with 300% torque is needed. To meet the need, trawling grounds have been shifting to deeper waters, the rope winding speed of winch is also to be as fast a 70 to 80 m/min at the mean diameter of its drum. This is why larger trawl winch is requested to day.

Table – 2: Relations between the maximum continuous shaft horse power of a main Engine P (ps) and the shaft horse power Tw (ps) of a trawl winch

Gross Tonnage	Max. continuous shaft HP of main eingine	Power to pull trawl EHP(ps)	Trawl winch Capacity ton x m/min	Shaft HP Tw
314	1200	130	8x80	142
540	1500	162	10x80	176
1500	2700	360	15x70	234
2800	3150	415	18.6x60	248
3000	3500	570	19.7 x 66	293
3400	4000	720	20x80	350

Table 2 and Fig. 44 show the relation between power to pull trawl computed in accordance with the preceeding paragraph and shaft horse power of trawl winch installed on the experimental trawlers.

From Fig. 44 it is clear that power to pull trawl (EHP) and shaft horse power of trawl winch (Tw) are nearly the same in the case of trawlers of 314-540 ton class (1200-1500 PS class), but with the increase in the size of trawler, its winch shaft horse power becomes slightly smaller than its power to pull trawl.

If a trawler's speed in hauling operation slows down to half of the usual towing velocity, the shaft horse power of trawl winch required to wind up warps at the same speed can be half of the power to pull trawl, theoretically. For example if towing velocity 4 knots drops down to 2 knots and trawl winch runs to wind up warp at a speed of same 2 knots (60 m/min), the speed of the gear to water is still kept at 4 knots and gear resistance does not change, accordingly. In the above case the warp winding speed of winch is half of towing speed, so the shaft horse power of the winch can be half of EHP.

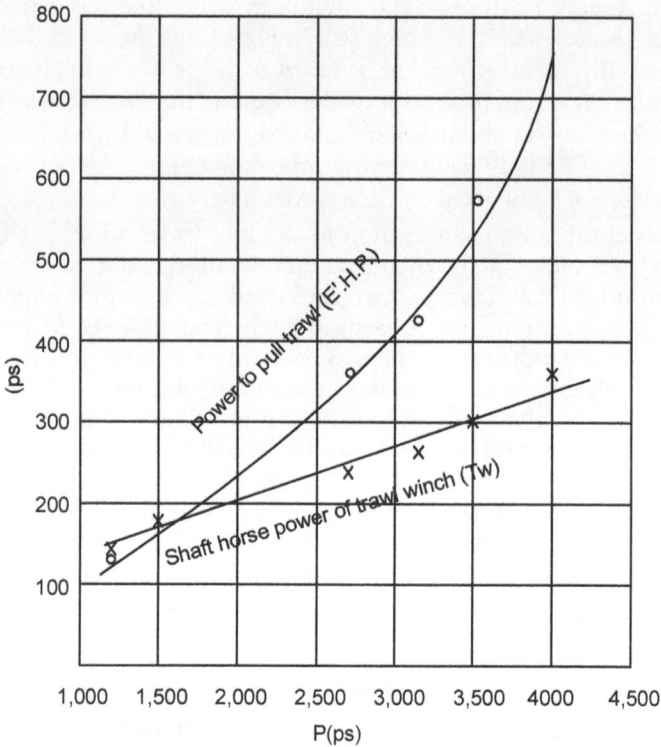

Fig. 44: Relations between the max. continuous shaft horse power of a main engine P (ps) and the shaft horse power Tw (ps) of a trawl winch and power to pull Trawl EHP (ps).

If a high speed winch, for instance 80 m/min, is requested, however, the shaft horse power of winch must be greater than half of EHP even though the towing speed is slowed down to half of the usual towing speed. It has to be nearly equivalent to the EHP, as observed in the case of 314-540 ton class. But it is not always possible to install a large traw! winch whose shaft horse power is as great as the EHP. There may be a problem of insufficient space on the deck. If such is the case, there is no other way but to control the winding speed of winch during hauling operation.

According to Fig. 44, the relation between the shaft horse power of winch (Tw) in use today and the horse power of main engine (P) is as follows :

$$Tw = 80 + 0.06P \qquad \qquad ...(2)$$

Trawl net : Even though trawl nets are equally designed and uniformly fabricated, they do not have uniform resistance if there is a difference in the buoy-ancy of their floats or in the opening breadth of their wings. It is very difficult, therefore, to accurately compute the resistance of a trawl net.

An empirical formula which is going to be introduced here is an approximate expression obtained from the results of experiments carried on ten kinds of trawl nets being used by 7 different trawlers of 100 gross tons with 300 PS engine to 3500 gross tons with 4000 PS engine. The experiments have been conducted under common and normal conditions, *i.e.*, the total buoyancy of floats is 200 to 650 kg, the total weight in water of ground bobbins greater than the total buoyancy of floats by 20 to 30%, netting is made of polyethylene twine and English knotted, towing velocity is 3.0 to 4.7 knots, and opening breadth of wings is kept in the range from 15 to 35 m. Fig. 45 shows the results of the experiments.

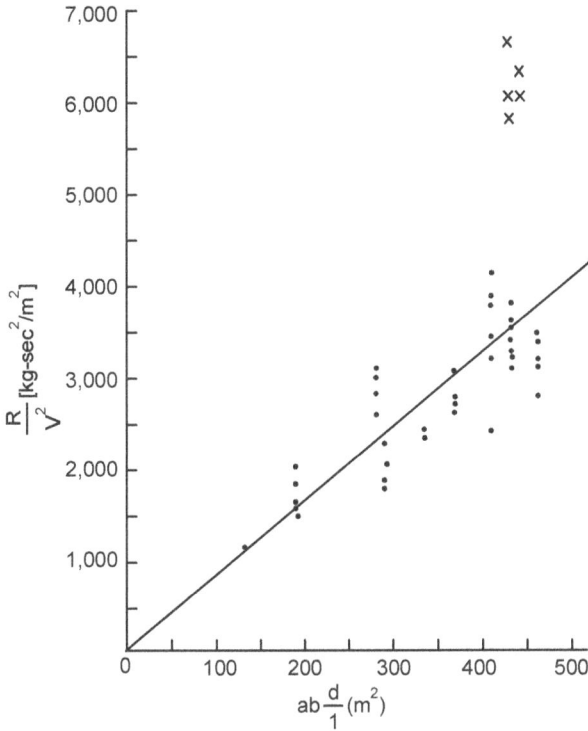

Fig. 45: Relationship between R/V square (kg.sec sq./m sq) and ab d/1 (m sq)

From Fig. 45 it can be said that the relation between R/V square (kg. sec sq/m sq) and ab d/1 (m sq) is almost proportional. The above abbreviations stand for:

R(kg) : resistance of trawl net; V (m/sec) : towing velocity

a (meter) the maximum breadth of body net without shortening (shown in dotted line in Fig. 46)

b (meter) :the maximum length of trawl net without shortening (shown by solid line in Fig. 46)

d/1 : in case of four seam net; the average value of ratio; d – diameter of net twine, I – length of each mesh bar at side panel, 1–7 sections in Fig. 46 in case of two seam net; the average value of ratio; d- diameter of net twine, 1 – length of each mesh bar at upper net, 1 – 7 sections in Fig. 46.

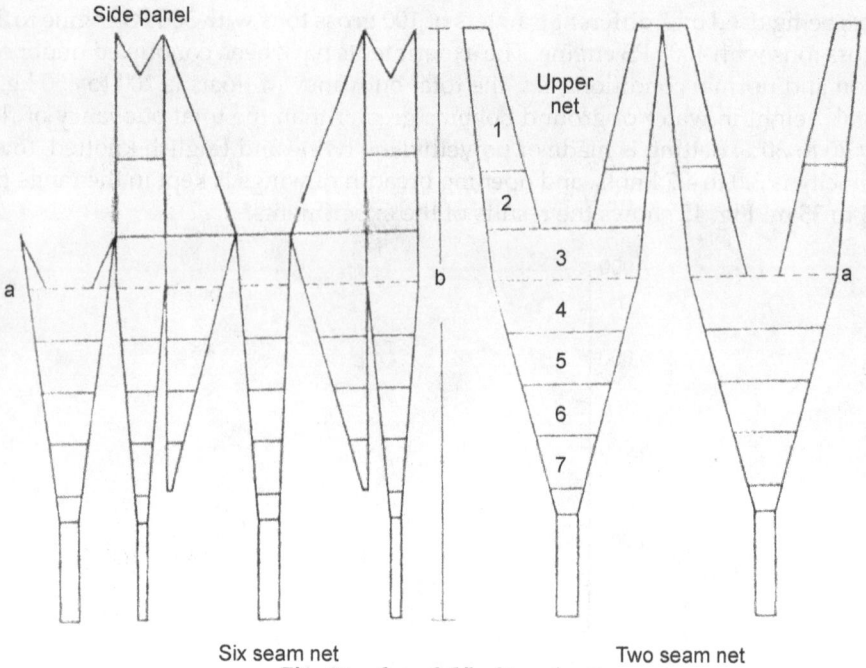

Six seam net   Two seam net

Fig. 46: a, b and d/l of trawl net

a; the maximum breadth of net body (dotted line)

b; the maximum length of trawl net without shortening (solid line)

d/l in case of four seam net;

>   the average value of ratio; d- diameter of net twine,

>   1- length of each mesh bar at side panel, 1-7 section in case of two seam net;

>   the average value of ratio; d- diameter of net twine,

>   I- length of each mesh bar at upper net, 1-7 section.

The d/l at cod head and cod end is not included in the above average ratio, because uniformity cannot be expected there due to the use of double twines in some trawl nets.

Then, from Fig. 45, the resistance of trawl net (R) with none fish catch is expressed approximately as follows;

$$R = 8\ a\ b\ d/I\ V\ square \qquad\qquad ...(3)$$

The above R is the resistance of the whole trawl net including floats, bobbins and other accessories as well as netting. The following is an example of computing a large trawl net resistance. The structure of the example net is shown in Fig. 47.

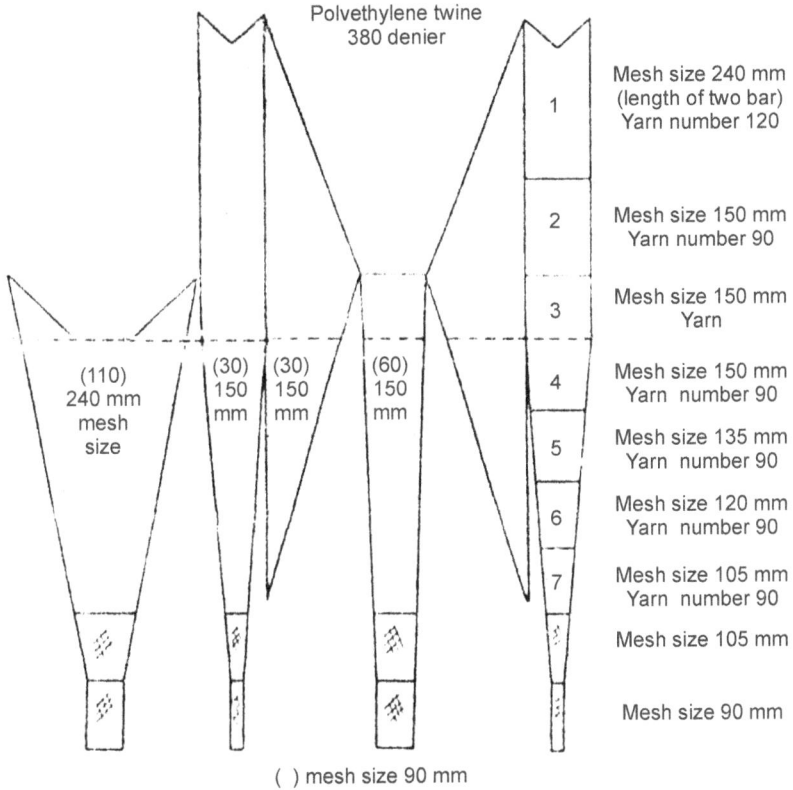

Fig. 47: Trawl net (example)

From Fig. 47,

$$a = 24\text{cmxll6}+15\text{cmxl40x2} + 15\text{cmx90} = 83.34\text{m}.$$

$$b = 24\text{cmx70}+15\text{cmx50}+15\text{cmx65}+$$
$$15\text{cmx65}+13.5\text{cm50} + 12 \text{ cm x } 50+ 10.5 \text{ cm x } 50+$$
$$10.5 \text{ cm x } 40+ 9 \text{ cm x } 120 = 76.8 \text{ m}.$$

The values of d/l of the respective sections are obtained from Fig. 47 and Fig. 48 are as follows;

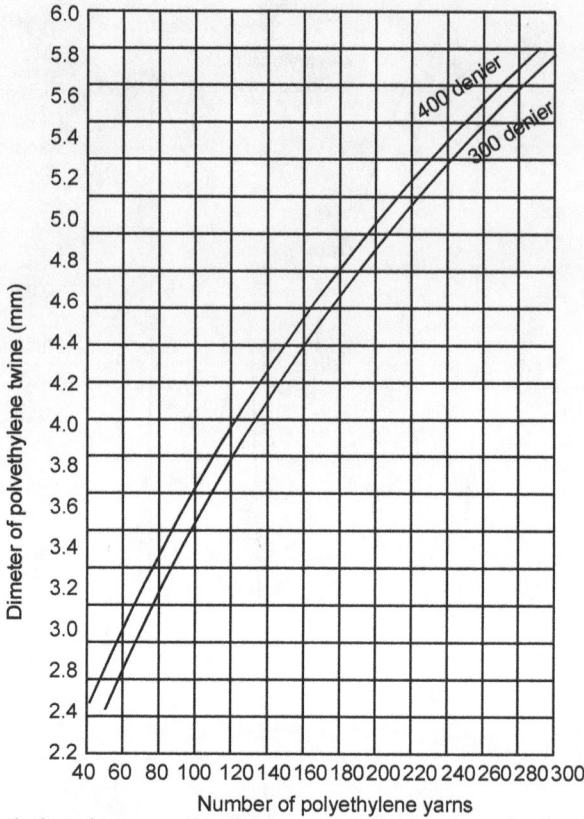

Fig. 48: Relations between the diameter and the number of polyethylene yarns.

Section 1 : 3.67/120 = 0.0306

Section 2 : 3.2/75 = 0.0427

Sections 3: 3.2/75 = 0.0427

Section; 4: 3.2/75 = 0.0427

Section 5 : 3.2/67.5 = 0.0474

Section 6 : 3.2/60 = 0.0534

Section 7 : 3.2/52.5 = 0.061

And the average value of the above figures is 0.0458

$$ab\ d/1 = 83.34 \times 76.8 \times 0.0458 = 292$$

Therefore, the trawl net resistance R = 8 x 292 V square. If towing velocity (V) is 4.5 knots (2.25 m /sec), R = 11.8 tons.

Since the old days, in Japan, the size of a trawl net has been indicated by the length of its head rope. The results of a survey on the relation between the main engine's horse power (P) and head rope length (H) of trawl net in use today are introduced in Table 3 and Fig. 49. According to Fig. 49, the relation between head rope length, H (m), and P (Ps) is expressed approximately by :

$$H = 42 + 0.006\ P \qquad \qquad ...(4)$$

However it is considered better to use the value of ab d/I (m square) when calculating the resistance of a trawl net.

Table – 3: Relations between the maximum continuous shaft horse power of a main Engine P (Ps) and the length of head rope of a trawl net H (m).

Gross tonnage	Max. continuous shaft horse power of a main engine P (Ps)	Length of head rope H (m)
314	1200	49.94
1500	2000	51.20
1500	2700	60.50
1500	2700	58.10
1850	3150	61.90
1850	3150	61.90
2800	3150	59.95
2800	3150	61.90
3000	3500	60.00
3000	3500	65.00

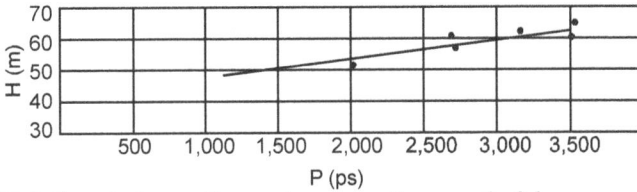

Fig. 49: Relations between the maximum continuous shaft horse power of a main engine P(Ps) and the length of head rope of a trawl netH(m).

Otter board : An upright type otter board, which resembles Suberkrub otter board shown in Fig. 50 is widely used in Japan. This type of otter board has less resistance and greater developing force than the plate type.

According to Scharfe, the relation between its developing force coefficients and resistance coefficients for different attack angles is shown in Fig. 51. The figure shows that the optimal angle of attack is 14-15 degrees and that when it is so, resistance coefficient is about 0.3.

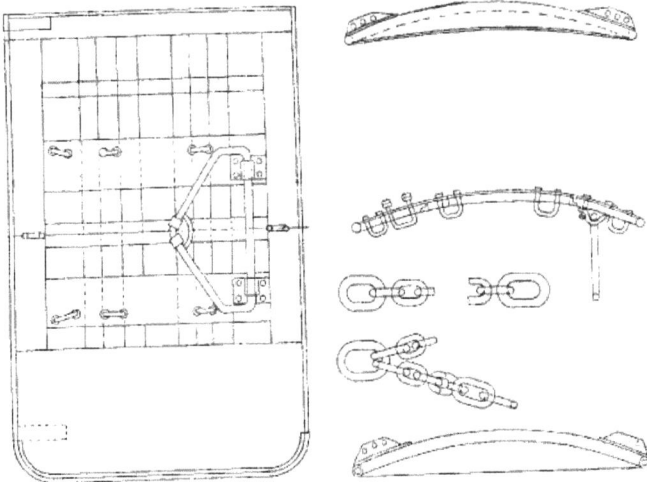

Fig. 50: Otter board used by 3150 ps type trawler.

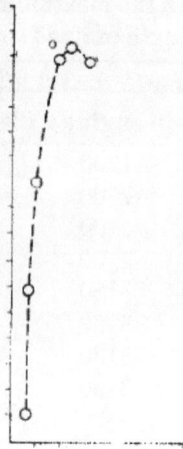

Fig. 51: Developing force coefficients (Ca) and resistance coefficients (Cd) for the Suberkrub otter board by Scharfe.

In the practical operation, otter boards are used under the condition where their developing force is kept at its maximum by means of adjusting the brackets angles. Therefore it may be safe to say that the attack angle of otter boards under operation is equal to the optimal angle, i.e., 14-15 degrees. This means that there will not be many errors if we use 0.3 as the resistance coefficient in the computation. Under this condition, the resistance of otter board (R" kg) is expressed as follows:

$$R'' = 1/2 \ CdpV \ square \ S \hspace{2cm} ...(5)$$

Where Cd is resistance coefficient, p is the density of sea water, that is 105 (kg.sec square/ m to the power 4), V is the towing velocity in m/sec and S is the area of otter board in square meter.

As an example, the resistance of a pair of Suberkrub otter boards of 3.8 meters in height and 2.5 meters in width may be computed as;

From the formula (5),

$$R'' = 1/2 \times 0.3 \times 105 \times 3.8 \times 2.5 \ V \ square = 150 \ V \ square.$$

Design features of trawls operating in Gulf of Mannar

The design features of trawls operating mid-way of the shoreline of Gulf of Mannar have undergone changes day by day. The modifications have been brought out based on trial and error methods, either by fishermen or by local net makers. The study conducted to find out whether the changes made in the trawl designs have any scientific base and to know how far the design features of trawls operated from this coast varied from standard FAO fish cum shrimp trawl (Pajot, 1987) recommended for operation from small class vessels of 32 to 39 feet OAL in Sri Lankan and Indian coastal waters.

Fifty trawl nets, each of 25 fish trawls and 25 coastal shrimp trawls operated from small mechanized vessels (36-38 feet OAL fitted with 68 to 88 HP engines) were studied for design characteristics, such as, mesh size, number of meshes, cutting rates, diameter of twine of various sections of fish and shrimp trawl. Nineteen parameters, namely, head rope length, foot rope length, length of cap, breadth of cap, length of top wing, length of bottom wing, breadth of top bosom, breadth of bottom bosom, length of overhang, length of first belly, breadth of first belly, length of second belly, breadth of second belly, length of cod end, breadth of cod end were considered. The measurements, such as, length of head rope and length of foot rope were taken to the nearest cm. The stretched length and breadth of each section were estimated by multiplying the number of meshes in length and breadth with the mesh size in the corresponding sections respectively. The morphometric measurements were expressed as percentage of Head rope length (HRL) for easy comparison and to work out percentage of overlapping of various parts of fish trawls and coastal shrimp trawls with each other and with that of a 26.6 m FAO fish cum shrimp trawl (Pajot, 1987). The percentages of overlapping of body proportions of various parts were worked out by methods of Hubbs and Hubbs (1953).

The design drawings of representative fish and shrimp trawls were prepared based on the stretched length and breadth of the webbing at various section of the net.

The range and mean values of seventeen morphometric measurements measured for fish and coastal shrimp trawls are given below.

Fish trawls: Length of head rope – 18-32 m (26.93 m), length of foot rope – 21–38 m (30.05 m), length of cap – 2.8-4.5 m (3.45 m), breadth of cap – 2.8-7.2 m (3.45 m), length of top wing – 7.8-13.8 m (10.5 m), breadth of top wing base – 8.0–10.8 m (9.45 m), length of bottom wing – 9.6-16.8 m (12.75 m), breadth of bottom wing base – 6.0–12.0 m (9.55 m), breadth of top bosom – 2.4–7.2 m (4.95 m), breadth of bottom bosom –3.6–8.4 m (6.46 m), length of overhang – 1.6–3.0 m (2.28 m), length of first belly – 7.2-18.0 m (10.65), breadth of first belly – 16.0-28.8 m (21.47 m), length of second belly –2.4-6.6 m (4.78 m), breadth of second belly – 6.0–19.2 m (12.43 m), length of cod end –5.4–8.2 m (7.23 m) and breadth of cod end – 2.0–3.0 m (2.42 m).

Coastal shrimp trawls: Length of head rope – 17.0-32.0 m (24.28 m), length of foot rope 19.0–36.0 m (27.20 m), length of cap – 2.8-4.0 m (3.22 m), breadth of cap – 2.56–5.40 m (4.04 m), length of top wing – 5.6–8.8 m (7.62 m), breadth of top wing base – 6.8–9.6 m (8.38 m), length of bottom wing – 7.2–12.9 m (9.99 m), breadth of bottom wing base – 6.0–8.8 m (7.8 m), breadth of top bosom– 4.8–7.2 m (6.02 m), breadth of bottom bosom – 5.4–10.4 m (7.44m), length of overhang– 1.60–2.72 m (2.16 m), length of first belly – 10.0-18.0 m (13.65 m), breadth of first belly – 6.5–21.5 m (16.4 m), length of second belly – 5.5–9.0 m (7.5 m), length of cod end – 6.4–9.1 m (7.46 m), breadth of cod end – 0.8–2.2 m (1.92m).

On comparing the percentage overlapping of sixteen body proportions of fish and shrimp trawls, it was found that the fish and coastal shrimp trawls varied from each other in five parameters with the overlapping less than 25%. The parameters by which the fish trawl and coastal shrimp trawls differed from each other were (a) length of bottom wing (14.12%), (b) breadth of top bosom (18.33%), (c) breadth of bottom bosom (18.05%), (d) breadth of second belly (21.94%) and (e) length of second belly (18.28%).

The reasons for relatively low percentage of overlapping of five body proportions of fish trawl with that of coastal shrimp trawl were due to excessive (i) bottom wing length (49.1% of HRL), (ii) top bosom breadth (19% of HRL) (iii) bottom bosom breadth (24.62% of HRL) (iv) second belly breadth (47.88% of HRL) and (v) second belly length (23.42% of HRL) in fish trawls.

A relatively higher bottom wing length in fish trawls than that of shrimp trawls could be attributed to overhang which was relatively longer in fish trawls than shrimp trawls. Keeping relatively broader bosom in the top and bottom panel in the fish trawl was found to be the practice of local fishermen as evident from relatively higher breadth of top bosom (19.0% of HRL) and bottom bosom (24.64% of HRL) in fish trawls. In general, the width of the mouth, which is decided mainly by breadth of belly in two seam trawls was found to be higher in fish trawl (59.20% of HRL) than shrimp trawls (40.86% of HRL).

The broader second belly in fish trawls was due to higher breadth in the first belly with the same cutting rate of 1P2B both in fish and shrimp trawls. The coastal shrimp trawls had the longer first belly in terms of percentage of head rope length (59.2%) against (40.86%) in fish trawls. The length of first belly in shrimp trawls may be relatively shorter than that of fish trawls as shrimps may escape through large meshes of first belly. As far as total of belly was concerned, the shrimp trawls had longer belly (73.60% of HRL) than fish trawls (64.28% of HRL). In general, the total belly length may be less in shrimp trawls and more in fish trawls as fishes have the more tendency to escape by moving against current while slowing the boat for hauling.

The fish trawls and coastal shrimp trawls when compared with the 26.6 m FAO fish-cum-shrimp trawl since all of them had almost similar head rope length (fish trawls and shrimp trawls were of 28 m and 24 m respectively), the deviations in the body proportions of the FAO fish cum shrimp trawl from that of local fish and coastal shrimp trawl is given below.

The body proportions by which the fish trawls deviated from FAO fish cum shrimp trawl exceeding 10% were (a) breadth of top bosom (11.49%), (b) breadth of bottom bosom (18.23%), (c) breadth of first belly (13.18%) and (d) length of first belly (13.80%). In all these four proportions, the fish trawls exceeded the FAO fish cum shrimp trawl stressing need to decrease proportion of these parts in fish trawls for their optimum performance. In general, the mouth of fish trawls of Gulf of Mannar region have been made wider by increasing the width of bosom and length of wing. The coastal shrimp trawls deviated from fish cum shrimp trawl of FAO with the deviation exceeding 10% in three morphometric proportions. These were (i) second belly breadths (11.23% of HRL), (ii) first belly length (32.14% of HRL) and (iii) second belly length (13.37% of HRL). Of these three proportions, the fish cum shrimp trawl of FAO exceeded the coastal shrimp trawl of Gulf of Mannar coast in breadth of second belly. A remarkably higher proportion of first and second belly in the coastal shrimp trawls of Gulf of Mannar than the fish cum shrimp trawl of FAO revealed that the belly length has been un-necessarily increased in the coastal shrimp trawls of Gulf of Mannar coast which would re-markably increase the drag resistance. In general, shrimps are weak swimmers and can not swim against the water flow as fish and hence increasing the belly length on par with fish trawl or more than that is not necessary. In fish cum shrimp trawl of FAO the proportion of first and second belly was found to be almost equal (27% of HRL). The proportion of overhang of fish trawls were higher than the proportion recommended by FAO and lower in case of shrimp trawls. Further, the length and breadth of cod end of both fish and coastal shrimp trawls were in accordance with the length and breadth of cod end of fish cum shrimp trawl of FAO.

The body proportion of the fish trawls and coastal shrimp trawls did not differ from each other and from that of FAO fish cum shrimp trawl design in parameters, such as, mesh size, twine diameter, hanging coefficient. The use of netting relatively with smaller meshes in the first belly of the shrimp trawl (50 mm) than that of FAO fish cum shrimp trawl (60 mm) would increase the drag resistance and prevent the escapement of shrimp juveniles. However, in the case of fish trawls, the usage of net panels with bigger meshes in the wing and overhang (120 mm) than that of FAO fish cum shrimp trawl (60 mm) is an encouraging feature as this would reduce the drag resistance and pave way for the escapement of juveniles of fishes. The usage of bigger meshes in the first belly of fish trawl (80 mm) is also an encouraging feature of fish trawls. Further, the cutting rate in the bottom wing may be adopted as 1P4B instead of 1P3B in both fish and coastal shrimp trawls which would help to increase the horizontal spread of the wing.

Pillai *et al* (1985) analyzed three different concepts of trawl designs and concluded that bulged belly trawl is more suitable for fishing prawns and either six seam or high opening trawl can be operated for capturing fishes. The fish trawls of Gulf of Mannar coast may be modified as four seam high opening trawl as it is found suitable for fishing fin fishes. Bulged belly trawl designs are suitable for fishing coastal shrimps.

Even though the absence of tickler chain in the shrimp trawls of Gulf of Mannar region is a technical flaw, fitting the tickler chain need not be recommended as this would lead to excessive disturbance of sea bottom which has been harbored with coral beds.

Irrespective of the type, the trawls had the cod end mesh size of 20 mm as against 40 mm in FAO fish cum shrimp trawl. Operation of 20 mm mesh cod end would pose over fishing of many commercially important fishes of the coast due to the contribution of considerable amount of juveniles in the commercial catch round the year. Therefore, square mesh panels of 30 mm mesh size may be fitted both in fish and shrimp trawls to facilitate escapement of juveniles.

3 | Working Principles and Performances of Trawl Nets

Catching methods and ways to improve them have been engaging the attention of fishermen from time immemorial. This was done mostly by trial and error methods, as most of the earlier investigations were primarily directed towards solution of biological problems related to fisheries. In recent years several fisheries laboratories have taken up studies on the working principles of many gears, such as trawls, gill nets, round haul nets etc with the aid of instruments developed for the purpose.

The fishing gear is a hydrodynamic system, the behavior of which is controlled by many hydro-dynamic factors, important ones among which known to affect the functioning of a trawl net being, (1) vertical opening, (2) horizontal opening, (3) depth of operation, (4) tension on the warp, (5) resistance to motion of the different parts, (6) angle attack of the boards, (7) tilts in the boards, (8) mesh shapes at different parts of the net, (9) water flow around and inside at different parts of the net, (10) tension at the cod-end due to catch in the net, (II) curvature at the foot ropes and head ropes, (12) magnitude and direction of the water current and (13) shape of the boat. By far the most reliable method of assessing the performance of slow moving systems, such as, trawls is direct observation. Specially trained divers (frog-men) are used in several countries for studying the working of trawls and similar gear. Under-water photography and underwater television are more recent methods. Since light is easily absorbed in water and is insufficient for illuminating objects which are not very close to the camera, some new developments have been made in under-water television replacing light by ultrasonic waves.

Instrumentation: Under-water instrumentation is comparatively more difficult because of the various restrictions in their design and operation. Weight, bulkiness, power source etc become very significant in the case of underwater measurements. The instruments attached to a net in particular should be very light and small in size so that the performance of the net is not affected by their presence. The instruments should be capable of withstanding rough handling. The power source required for the ultrasonic type transmitting instruments is yet another problem, in that the power consumption has to be kept at a minimum. Although many advanced and efficient techniques suitable for other fields are known, these can not be adopted in under-water instruments and hence special methods are required to be evolved.

The instruments so far developed include (1) mechanical under-water recording type, (2) wire telemetering and (3) wireless ultrasonic telemetering types. In addition to their being heavy and bulky, the main drawback of the underwater recording instruments is that the information can be obtained only after hauling the gear. The wire telemetering types possess the advantage that their sensors fitted on net are small and light and hence several of them can be used at a time. But the major handicap is the difficulty in handling the cable. The wireless ultrasonic transmitting type instruments are of recent origin and are easy to operate. They are relatively heavier than wire telemetering types. Further there are apparent difficulties in measuring most of the parameters of the fishing gear in this method and hence their use is limited only to a few of the above said hydrodynamic parameters. Moreover, the echoes of the ultrasonic signals from the bottom, water surface etc cause errors in the measurement and are still unsolved problems in under-water wireless communication.

Telemetering instruments for instantaneous measurements.

Telemetering instruments provide instantaneous measurements of the parameters. Ultrasonic waves are the only wireless means of communication under water as electro-magnetic waves and light waves are easily absorbed in sea water. Ultrasonic waves have the advantage that they can travel longer distances in water at a speed five times that in air. But the production of ultrasonic waves is much more difficult compared to production of electro-magnetic waves and hence communication under water is more difficult than in air. The range of transmission of ultrasonic waves is very much limited (a few kilometers), while that of electro-magnetic waves in air and space can easily travel several thousands of kilometers. During the last decade several wire and wireless lelemetering instruments have been developed. The contributions made by the Marine Laboratory, Aberdeen, White Fish Authority, U.K., Fisheries Research Board of Canada, National Institute of Oceanography, Survey, Fishing Boat Laboratory, Japan, Bureau of Commercial Fisheries, U.S.A. and West Land Air Craft Ltd, U.K. are worth mentioning. Recently in India, Central Institute Fisheries Technology has developed wire telemetering instruments for the measurements of depth of operation, angle of attack of otter boards, tilt of otter boards, fore and aft tilt of otter boards, underwater tension for the measurement of the resistance to motion of the different parts of the net during operation, mesh size variations and water flow around and inside the net.Commercial firms like Furno Electric Co, Japan, Kode Electronics Ltd, Japan, Titran Electronics Co, Japan etc have developed and marketed ultrasonic equipments for the instantaneous measurement of the depth of operation, vertical opening of trawls, recording and counting of fishes entering the net etc.

Continuous multi-channel data acquisition systems.

Many of the hydrodynamic parameters are inter-dependent and it will be more valuable to obtain at least the most important of them simultaneously and instantaneously. Telemetering instrumentation is the only way of satisfying these conditions and many fishing research institutions all over the world are engaged in the development of multi-channel data acquisition systems. Foster (1966) described a fully instrumented gear for measuring the important parameters of the trawl. White Fish Authority, U.K., in cooperation with a commercial firm has developed a multi-channel ultrasonic link between the net and the boat. This system measures (i) water temperature at the sea bed, (ii) head line height (vertical opening), (iii) net mouth spread (horizontal opening) and (iv) strain in the head line legs of the trawl (under water tension). The system is also provided with a calibration channel for checking the normal working of the transmitter. The details of the working system are described elsewhere by Hearn (1966). Mowat (1966) describes the working details of a digital acoustic telemeter for fishing gear research.

The data from the various telemetering units received in the form of continuous analogous or digital forms can be displayed in a meter or stored in either continuous paper recorders or tape recorders. These data are processed for obtaining the best working conditions of the trawls. Johr (1966) describes the attempt made in Norway in the wireless data acquisition system in fishing gear research. They have succeeded in measuring the tensile loads in the warp before and after the otter board, depth of the otter board, the three position angles of the otter board, distance between the two otter boards and net opening. The data are converted to electrical signals by suitable transducers. They are time multiplied and sent to an analogue-to-pulse frequency converter, the output pulses of which control the transmitter of the ultrasonic telemetry link.

Data processing with the help of computer.

Data processing is equally important as instrumentation. The data are processed both for studying the effect of each individual part as well as for deciding the performance of the fishing gear under the various known and diverse circumstances. Computers are used for quick analysis of the data from the several instruments used. The data already obtained from the various instruments working on different principles are to be converted to a uniform digital form for feeding the computer, which is achieved by using analogue-digital electronic converters. The marine laboratory in Aberdeen used an electronic computer for gear research works especially developed by Elliot Bros, as per the specific requirements (Anon, 1970). This computer contains a "data logging program" of 6000 separate instructions for the control system. Information from 16 instruments on the net were first brought to a telemeter control unit on the net and they are transmitted to the boat ultrasonically. Such types of computers are now in use in many other fields of marine engineering especially in automatically controlling the functions of the ship, navigation and also in designing the ships.

Trawl Net Design and Development

Development and trial of 25 m. bulged belly and six seam trawl net

A bulged belly trawl with 25 m head rope and a six seam trawl of 25 m head rope were operated together with flat rectangular otter boards of 1524 mm x 762 mm size weighing 100 kg each in air. Fishing was conducted in 1977 from a 15.2 m vessel fitted with 165 HP engine. The depth of the fishing ground varied from 26 to 55 m off Veraval, Gujrat. The details of two nets are shown in the figure 52. Both the nets were operated the same day keeping depth, ground, length of warp, trawling speed, duration of each haul and the course constant. Each net was provided with double sweep line of 5 m long between the leg and otter board.

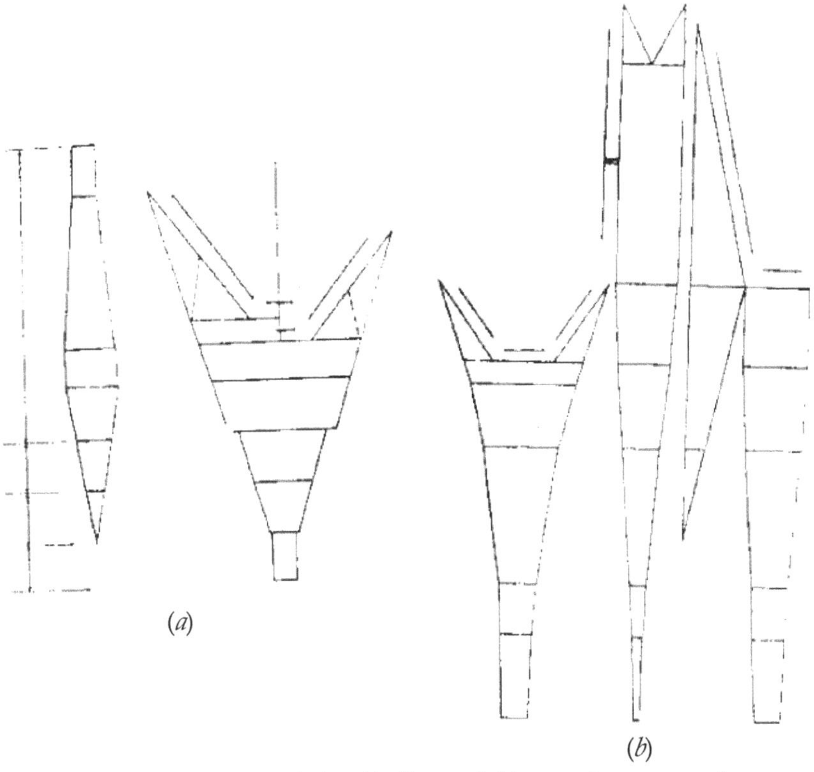

(a)

(b)

Fig. 52: (a) 25 m bulged belly trawl (b) 25 m six seam trawl

Details of 25 m six seam bulged belly trawl net.

Webbing	A	B	C	D	E	F	G	H	I	J	K	L	M	N	O	P	Q	R	S
Twine dia(mm)						15													
Breaking strength(kg)						36													

Stretched Mesh(mm)	63	63	63	63	63	63	76	76	63	63	63	50	50	50	50	38	38	31	31
Upper edge Meshes	19	01	95	79	90	11	44	15	07	09	51	06	25	67	132	88		44	96	28
Lower edge Meshes	79	90	45	95	21	78	35	124	105	54	84	66	1	33	66	80	24	80	16
Depth meshes	190	155	45	57.5	115	57.5	57.5	124	35	7.5	57.5	120	80	120	120	60	60	120

Baiting rate (inner) 1:4

Baiting rate (outer) 1:2.4-1:1-1:2 !;4.8 1:1 1:1.2 1:1.9 1:2.4 1:5 1:6 1:3.3 1:7 1:3.6 1:1 1;5 1:61:15 1:20

Hanging coefficient 0.87 0.87 0.87 0.50   -   -   0.97 0.50

Hanging a/AC= 11.0/12.6   c/BC= 11.0/12.6 - b/D = 3.0/6.0 −dg=3.5/3.6 c/H = 2.8/5.6

Blue high density polyethylene with single trawl knot was used. Total weight of the net -45kg.

The tension and horizontal opening were more in bulged belly and is more efficient than the six seam trawl with respect to catch. Bulged belly trawl landed more from the two depth zones (26-39 m and 40-55 m) compared to six seam trawl . Same trend was observed in deeper waters (beyond 40 m). The increase in catch of six seam net in deeper waters was due to the increased landings of ribbon fishes at random, whereas all the other fishes were caught at par with that of bulged belly trawl. It may be calculated that the bulged belly trawl may be more efficient in catching the "shallow water mix" consisting mainly of smalt miscellaneous fishes and crustaceans which are more abounded there than in deeper waters.

The percentage of horizontal opening of the two nets differ significantly. Six seam trawl had higher percentage of opening compared to bulged belly irawl at all depths. Reduced horizontal spread of bulged belly trawl may be due to increased bottom friction. Factors, like smaller mesh size, many meshes, and larger quantity of twine in bulged belly trawl would have offered more resistance which in turn might have reduced the horizontal spread. This was evident in shallow waters where bottom friction was more. However, there was no corresponding increase of catch in relation to the increased horizontal opening of six seam trawl in shallow waters. It may be noted that the increased horizontal opening obtained by six seam trawl had no added advantage over the bulged belly trawl in shallow waters.

Towing warp tension and horizontal opening between otter boards were recorded. The catch composition of each haul was recorded separately for the two gears, 74 hauls, each of one hour duration with both nets were made. The nets landed a total of 13659 kg of fish.

Rigging

Details of lines and ropes of bulged belly trawl

	a	b	c	d
Material	High density polyethylene			
Diameter in mm	18	18	18	18
Breaking strength(kg)	34.60	34.60	34.60	34.60
Length(m)	11.0	3.0	2.8	3.5

Head rope–25 m ;   Foot rope–31.8m.

Details of floats, sinkers and otter boards in bulged belly trawl

	Floats	Sinkers	Otter boards
Number	11	-	2
Material	Hard plastic	Iron	Iron and wood
Shape	Spherical	Link chain	Rectangular, flat
Diameter (mm)	156	6	-
Length (mm)	-	-	1524
Breadth (mm)	-	-	762
Static buoyancy (kg)	1.550 each	-	-
Weight in air (kg)	0.300 each	30.0	100.0 each

Details of 25 m bulged belly trawl

Webbing	A	B	C	D	E	F	G	H	I	J	K	L	M	N
Twine dia.(mm)						1.5								
Breaking strength(kg)						36								
Stretched mesh(mm)	<— — — — — —50——----->							40	40	30	30	20		
Upper edge meshes	60	1	435	1	60	120	410	120	350	110	270	80	225	150
Lower edge meshes	60	140	410	130	120	120	350	90	270	60	170	1	125	150
Depth meshes	60	140	280	25	130	260	165	45	45	60	60	75	75	100	100	150
Baiting-rate	1:2	1:2	1:2	1:5.5	-	1:1.5	1:41:1.5	1:3	1:1.5	1:2.5	1:2	-		
Cutting rate	– Ip2b	Ip2b	Ip2b	5p2b	-	Ip4b	3p2b	Ip4b	Iplb	Ip4b	3p4b	Ip4b	-	
Coefficient of hanging	0.75	0.60	0.45	0.75	0.90	0.58	0.47							

Hanging   a/A=2.25/3.00, b/B-8.50/14.00, c/C-3.50/7.75 d/A=2.25/3.00, c/E=1.50/1.65 f/D=7.5/13.00, g/G-3.50/7.50

Blue high density polyethylene with single trawl knot was used. Total weight of the net was 57 kg.

Details of lines and ropes

	a	b	c	d	e	f	g	leg	leg
Material	High density polyethylene								
Diameter in mm	<— — — — — — 18— — — — — — — — —-—>								
Breaking strength (kg)	«— — — — —34.60— — — — —---------------->								
Length (m)	2.25	8.50	3.50	2.25	1.50	7.50	3.50	5.00	5.00

Head rope – 28 m.   Foot rope – 26 m

Details of floats, sinker and otter boards

	Floats	Sinkers	Otter boards
Number	11	-	1 pair
Material	Plastic (hard)	Iron	Iron and wood
Shape	Spherical	Link chain	Rectangular, flat
Diameter (mm)	150	6	-
Length x breadth(rnm)-		-	1524x762
Static buoyancy (kg)1 .550 each		-	-
Weight in air (kg)	0.300 each	30.00	100 each

Comparative details of the two nets

Particulars	25 m bulged belly trawl	25 m six seam trawl
Total number of meshes	333000	172000
Range of mesh size (mm)	20-50	30-75
Quantity of twine (kg)	40	25
Quantity of ropes and lines (kg)	17	20

Results of comparative fishing with 25 m bulged belly and 25 m six seam trawls

	26-39 m depth		40-55 m depth	
Particulars	Bulged belly	Six seam	Bulged belly	Six seam
No. of hauls	57	57	17	17
Duration(hours)	57	57	17	17
Trawl warp tension(kg)				
Average	522.6	482.9	542.0	521.0
Range	448-658	403-605	425-684	425-630
Horizontal opening of otter boards in m.				
Average	18.83 (41.34%)	22.80 (50.78%)	25.87 (57.50%)	29.68 (66%)
Range	15.52-24.73	17.78-28.67	23.50-28.50	28.12-30.98
	(34.54-54.95%)	(39.63-63.72%)	(52.22-63.33%)	(62.48-68.80%)
Total catch(kg)	5679.55	4591.35	1362.45	2025.30
Catch/hour (kg)	99.64	80.55	80.14	119.13

The trawl warp tension of bulged belly trawl was more compared to six seam trawl at all depths. The bulged belly trawl was offering more resistance as indicated by the increased warp tension.

The catching efficiency of bulged belly trawl was found to be more in shallow waters; it can be recommended for the exploitation of shallow water mix. Both the nets were found equally effective for the capture of all other varieties of fish.

Since the six seam trawl was cheaper (32.4%) and with lesser warp tension (7.6% in shallow and 4% in deeper waters) than bulged belly trawl, the possibility of either six seam trawl to be increased in size or to be towed at a faster speed with the given engine power.

Large mesh high opening fish trawl

Large mesh high opening trawl is a recent introduction (Buckingham, 1972) and found very successful and efficient. The present trend in trawl design is to use larger meshes, as the schooling fishes seldom attempt to escape through large meshes of the fore part of the net. The increase in mesh size has greatly improved the catching efficiency of the net.

A large sized mesh trawl with 32 m head rope and high opening was designed and fabricated as specified in the figure and table. The mesh size has been increased from 50% in the cod end to 100% through the fore part of the net when compared to the conventional trawl. Because of the increase in mesh size the total number of meshes has been reduced to 120% and webbing to 35% compared to a 32 m bulged belly trawl described by Pillai *et al* (1978).

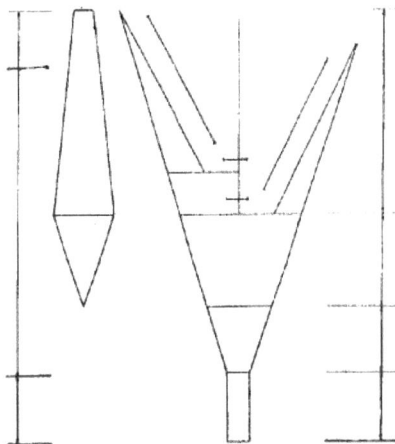

Fig. 53 : 32 m large meshed high opening trawl

Fishing was carried out from a 15.2 m OAL vessel, fitted with 165 HP Cummins engine during two fishing seasons at depths of 27 to 37 meter off Cochin. The large meshed trawl was operated together with a 32 m bulged belly trawl to compare the effect of increased mesh size. Otter boards of 1900 x 900 mm flat rectangular wooden board have been used. The nets were operated one after another keeping depth, length of rope, trawling speed and duration of haul constant. The quantity of different varieties of fish caught by both the nets were recorded separately. The horizontal opening between otter boards was measured by an angle measurer and the warp tension by an electronic warp load tension meter.

Details of 32 m large meshed high opening trawl

Webbing	A	B	C	D	E	F	G	H
Stretched mesh(mm)	120	120	120	80	120	80	40	30
Upper edge meshes	1	223	1	290	30	140	300	130
Lower edge meshes	48	193	48	150	90	1	100	130
Depth meshes	120	30	120	105	150	105	150	150

contd...

No. of pieces		2	1	2	2	2	2	2	2
Cutting rate		Ip3b	Ip2b	Ip3b	Ip4b	2plb	Ip4b	Ip4b	Allp
Baiting rate		-	1:2	-	1:1.5	1:5	1:1.5	1:1.5	-
Rope length (m)		a 12.5	b7.0	c12.5	d6.0	e3.0			
Hanging coefficient		87	45	87	50	80			

Particulars of large meshed trawl and bulged belly trawl.

Particulars	Large meshed high opening trawl	Bulged belly trawl
Head rope length (m)	32	32
Foot rope length (m)	37	35
Wing mesh size (mm)	150	80
Jib mesh size (mm)	120	60
Overhang mesh size (mm)	120	60
Body mesh size (mm)	80	50
	40	40
Cod end mesh size (mm)	30	30
Number of meshes	196000	410000
Twine diameter (mm)	1 and 1.5	1 and 1.5
Material required (kg)	18.5	29

Average horizontal spread between otter boards and average tension for two nets

Net	Depth (m)	Av. Speed (knots)	rpm	Av. Tension (kg)	Av. opening
Large meshed	27-37	3	1200-1250	450	30 m
Bulged belly	27-37	3	1300-1350	550	28 m

Catch composition and quantity offish caught in large meshed and bulged belly net Fishing season - 1977-78; Depth of operation - Up to 30 m; Trawling hours –33

Net type	Prawn	Squid	Quality fish	Barracuda	Cat fish	Sharks	Sciaenids	Total
Large meshed	16	-	19	21	56	37	2706	2855
Bulged belly	45	-	8	12	25	20	2620	2730
Fishing season-	1978-79; Trawling hours- 63; Depth - Beyond 30 m							
Large meshed	0	580	31	164	25	1131	7954	9885
Bulged belly	3	300	40	61	15	606	5780	6775

(All catches mentioned above are in kg)

Large mesh high opening trawl was found to be very efficient for the exploitation of column and off bottom fishes. The net was found more efficient than the bulged belly trawl in capturing different varieties of column fishes comprising of squids, pompfrets, barracudas, catfishes, sharks and rays. The net offers 18% lesser resistance which in turn results in utilization of lesser horse power. The trawl is very simple in construction and requires minimum maintenance because of reduced number of meshes. The total cost of the net is less by 35% as the larger mesh size reduces the material requirement and cost of fabrication.

Belly depth studies of shrimp trawls

A 13.69 m (45 feet) four seam cotton trawl net with 85 meshes in depth of belly was selected as the control net. Two other trial nets, 70 and 55 meshes in depth of belly were constructed reducing in all 30 meshes in the depth of belly in two stages. The dimension of the bellies of the control net as well as other two trial nets are shown in the figures.

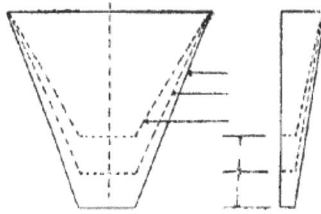

Fig. 54: Control net (85 meshes in depth)

(a) Trial net (70 meshes in depth)   (b) Trial net (55 meshes in depth)

On each day of operation, all the three nets were operated in rotation and as far as possible all the fishing conditions, direction of tow, number of floats on head rope, weight of sinkers on foot rope and size of otter boards were kept constant.

Nets	No. of hauls	%opening	Tension (kg)	Catch in kg		
				Prawn	Fish	Total
Control (85 meshes in depth)	24	47.60	304.0	7.02	13.60	20.62
Trial net (70 meshes in depth)	24	48,20	303.3	11 .64	18.20	29.84
Trial net (55 meshes in depth)	24	47.07	305.5	7.90	15.16	23.06

The fishing results showed that net with 70 meshes has an optimum depth of belly as compared with 85 meshes and 55 meshed in depth of bellies. The reduction of belly depth beyond 70 meshes may be detrimental to the catching efficiency as well as the mechanical characteristics as exemplified in net of 55 meshes in depth.

Six seam otter trawl

Trawling for capture of demersal fishes is now a popular fishing method in most maritime states of India and several designs of otter trawls are used. The nets presently used are basically either two seam or four seam in construction. In countries like Japan nets with more than four panels are fairly popular.

With a view to assess the catching efficiency of such a design, a 15.8 m six seam nylon trawl net was fabricated for conducting field trials.

The design and construction details of 15.8 m net along with other accessories used in the operation are given in the figure and table below. The otter boards used for the net were similar to that used in four seam and two seam tawl.

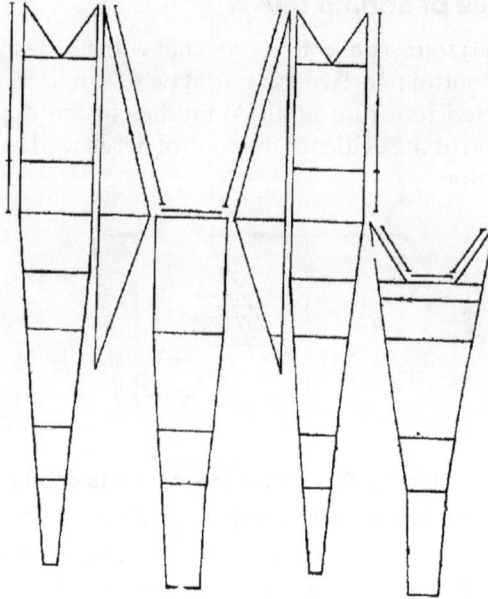

Fig. 55: Construction details of a 15.8m six-seam trawl

Particulars of gear accessories

Head rope – Manila, 15.9 m, 12.7 mm dia.; Foot rope – 20.5 m, 19.1 mm dia.; Sweeps – Head rope – Manila, 11.8 m, 12.7 mm dia.; Foot rope – Manila 11.8 m, dia. 19.1 mm; No, of floats – 15; Total extra buoyancy of floats attached – 12.5 kg; Total weight attached along the foot rope –27 kg (1.3 kg/m).

Otter boards – Rectangular flat type, Size – 1.4m x 0.63m, weight - 55 kg each.

Towing warp – 12.5 mm dia., galvanized, flexible steel wire rope.

With a total 100 hauls in 14 days for 75 hours duration, 2337.3 kg of prawns and 15728 kg offish were caught with the six seam trawl net; the average catch per towing hour being 30.56 kg for prawns and 209.79 kg for fish. The species caught consists of prawn, lobsters, Sciaenids, Polynemids, Eels, Elasmobranchs, Perches, Pomfrets, Ribbon fish, Silver bar, *Pellona* sp and miscellaneous varieties.

Except pomfret and *Pellona* sp, the other varieties landed by six seam trawl were same when compared with those landed by two seam and four seam trawls.

The average horizontal opening between otter boards during operation came to 22.6 meters, which works out to 57,3% of the buoy-line including sweeps. The towing resistance offered by the gear on both the warps was observed to be 5.6 kg only.

Three panel double trawl net

Two separate nets mounted one above the other, a horizontal panel was introduced in between the top and bottom belly of a two panel net. The design details and construction particulars of the trawl net are shown in the figure.

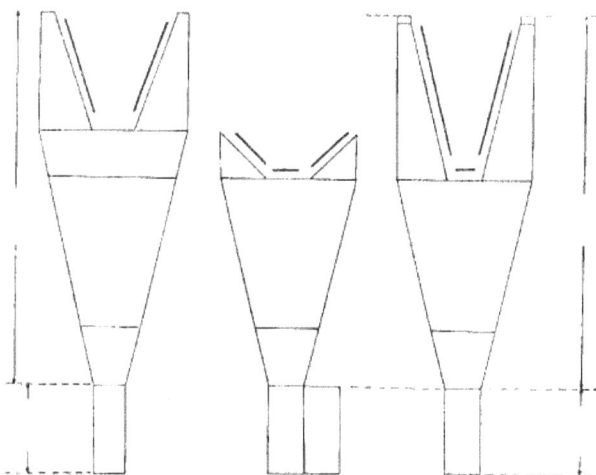

Fig. 56: Design of 10.97 m three panel double trawl net

The net essentially consists of a normal two seam type with a middle panel inserted to divide the net horizontally and provided with two separate cod ends. The rope attached to the central webbing and wooden floats was joined with lateral seaming lines of the net.

Details and results of operations of 10.97 m three panel double trawl

	Boat I	Boat II
Depth of fishing grounds in m	8 – 14	15 – 40
Warp length used in m.	50–80	80 – 210
Horizontal distance between otter boards(m)	7.53	22.09
Average/range	(6.17–8.47)	(18.62–26.67)
Towing tension in kg, Average/range	359.0	455
	(320-394)	(428-484)
Towing speed in knots, Average/range	2.0	2.16
	(1.50-2.30)	(1.70-2.37)
Number of hauls	25	20
Tow time in hours	24.00	19.30
Total catch in kg (in upper cod end)	50.50	260.25
Total catch in kg (in lower cod end)	273.50	615.75
Catch/haul/trawling hour in kg in upper cod end	2.10	13.35
Catch/haul/trawling hour(kg) in lower cod end	11.40	31.50
"(range) in upper cod end	(0.5-7.0)	(1.0-78.0)
"(range) lower cod end	(0.5-33.0)	(2.5-121)

Fishing trial was made in two series of varying depths and from two different boats of 9.13 m with 36 HP engine in the depth range of 8-14 m during October, 1969 and from 12.16 m with 60 HP engine in the depth range of 15-40 m during

May, 1971. Vertical curved otter board, dimensions 100 x 50 cm and weighing 35 kg were used in the case of fishing operations with the former boat; while V - shaped steel otter boards of dimensions 122 x 72 cm and weighing 55 kg were used in the case of latter. The net was directly attached to the otter board by its extension legs of 7 m length, when operated from the first boat; while in the second boat beside legs, single sweep wire of 20 m length on either side was provided. Trial fishing was conducted in the known fishing grounds.

The catch rate was better when the net was operated from the second boat and lower cod end yielded more fish than the upper cod end and the catch in the lower cod end was approximately 2.5 times more than the upper one.

More of bottom fishes like prawn, sole and sciaenids were found in the lower cod end; while off-bottom fishes like *Lactarius, Caranx,* pomfret in the offshore operations of second boat and ribbon fish in the inshore operation of first boat were more in the upper cod end. Nakamura (1970) described the utility of the two floor trawl net (vertical twin body net) in catching fish, such as, sole swimming close to the sea bottom with the lower net and such fish as herring swimming off the sea bottom with the upper net. The present trial revealed the effectiveness of this two tier net for catching the bottom and off-bottom fish simultaneously and its utility in separating them in the fishing operation itself.

The horizontal spread between the otter boards was found to be 30.12% and 34.14% of the head rope length of the gear when operated from the first boat in inshore waters and second boat in deeper depths respectively. This horizontal spread although low indicated the possibility of good vertical opening of the gear. This is further supported by the catch of off-bottom fishes in the upper cod end. The under water resistance of the gear, as measured on the towing warps showed reasonable difference between the two boats in view of the difference in the area and depths of operations, difference in the otter boards used and towing speed. In light of these findings, the trawl net described is very useful for the simultaneous and at the same time, separate capture of bottom as well as off-bottom fishes and can also be advantageously used in testing and exploring new grounds for the availability of different kinds of fishes.

Overhang for trawls: Certain species of fish try to escape from an advancing trawl by swimming upwards and forwards. The catches of fish trawls can therefore, be significantly enhanced by increase in head line height of the net. Many workers (Takayama *et al,* 1958 & 1959, Phillips, 1959, and Ben Yami, 1959) have adopted different techniques to obtain this vertical opening. Verghese *et al* (1968) made a preliminary attempt to increase the fishing height by redesigning the conventional trawl net into a bulged belly type. In view of the probable behavior pattern of some of the column fishes, it is likely that an improvement in the design of the bulged belly net can be brought about by adding an overhang, thereby minimizing the upward escapement of fish.

Nair and George (1964) based on a survey of existing designs, suggested certain relationship between the square (overhang) and the upper belly in shrimp trawls. However, in the case of fish trawls, information as to the optimum size of overhang is lacking.

The design details of the nets are given in figures 57A, 58A and 58B. The net 57A without square was the control net, and nets 58A and 58B with squares of length 30 and 60 meshes respectively were used for trials. All the nets were rigged with sweep wires of 20 m length and horizontal curved otter boards of size 120 cm x 60 cm. Trawling was conducted from 9.85 m boat with 40 HP engine in the sea off Cochin at depths ranging from 15 to 30 meters. Forty six fishing trials were undertaken from October, 1967 to April, 1968. Three nets were towed for one hour each per day and changed in regular rotation at successive hauls giving all the nets equal chance. Catch data were analyzed separately for prawn, fish and total catch.

The analysis of catch data indicate that the net 58A with a square of 30 meshes recorded better catches than the other two nets. Of the total catches landed by the three nets, 41.3% was landed by net 58A, 36.2% by net 57A and 22.5% by net 58B, indicating that net 58A might have engulfed maximum number of fishes that have tried to escape upwards.The lengthy square 58B on the other hand, recorded minimum catch due to increase in length of square, which might have caused the head line to sag and consequently decreased the fishing height.

The size of the overhang has a bearing on the yield of the net and optimum size of the square derived at is 1.9 meter.

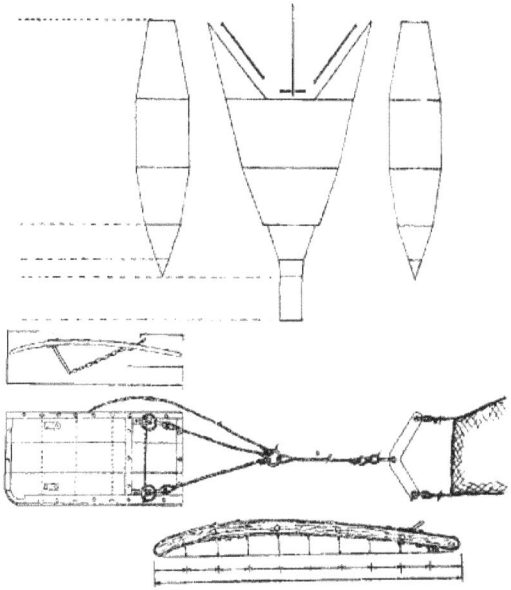

Fig. 57: A Design diagram of trawl net without square (Control net)

58 • A

58 • B

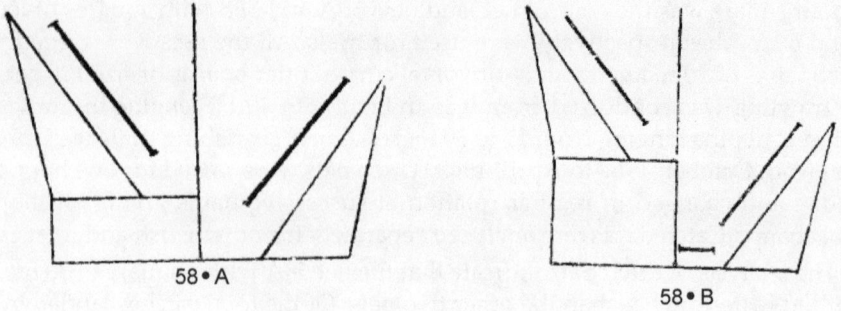

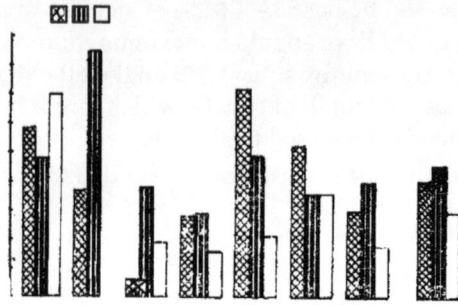

Fig 58: A and B Design of trawl nets of 30 meshes and
60 meshes square length

The mid water trawl was first designed and operated by Robert Larsen of Denmark in 1948. His net was a four seam equal panel square net operated by two boats. Later considerable success was made in Germany with one-boat mid water trawl of rectangular shape and having lower panel bigger than the upper one.

Features: Mid water trawl designs are characterized by large vertical and horizontal opening and with smooth water flow inside the net. The flow characteristics are particularly important to reduce the turbulence at high towing speeds. The smooth catenery construction of head line and foot rope with long finely tapered bag are incorporated in successful design.

Babylon : The smooth catenary on trawl mouth is made by the construction of hang meshes at the quarter shoulder called "Babylon". The term "Babylon" is adopted by Norwegian fishing gear factories. In Fig. 59 the dotted line shows the position of trawl mouth without Babylon which reveals the necessity to make framing lines nearer to to fishing shape. The hang meshes at the quarter shoulder are fabricated with double twine.

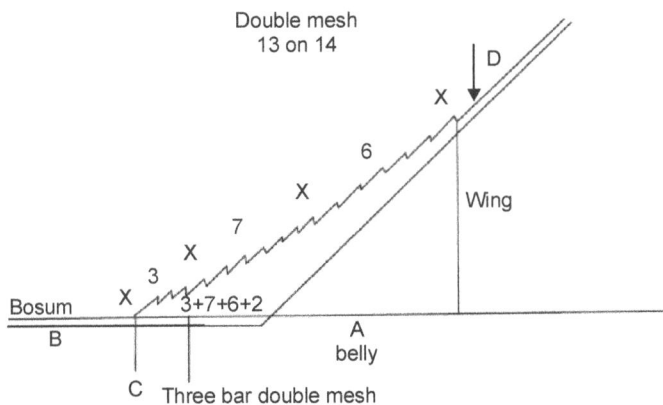

Fig. 59: Catenary when Babylon is constructed

Construction : The all bar wing is joined at A to the belly with double mesh. From bosom of the belly, one mesh (2 bars) is cut at B and double meshes are made instead with the same gauge. Now to reach the joining lines of bosom and wing 3 bars have to be fabricated, i.e., 2 cut bars and one joining bar.

Next step is the construction of Babylon. The number of Babylon for each design may vary and is arrived after trial and error to suit the catenery shape. The construction of Babylon 3-7-6 in design 6 x 6F squadratic mid water trawl (Fig. 60) is done in the following way. When 3 meshes are added to this, the base meshes will be 19. Out of 3 meshes, two are for wing and double meshes and one for double knot at the beginning of Babylon (C).

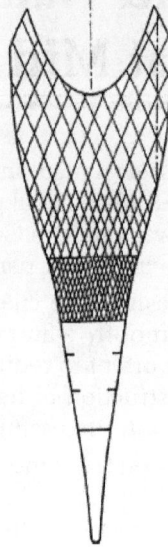

Fig. 60 : 6 × 6F Quadratic Mid water trawl

Babylon 3-7-6 means construction of 3 hang meshes at an interval of one bar (one bar Babylon equals to IP IB), next 7 hang meshes at an interval of 2 bars (2 bar Babylon equals to IP 2B) and last 6 meshes at an interval of 4 bars (4 bar Babylon equal to IP 4B) to form a smooth catenery at the corner of the bosom.

Before fabricating the Babylon, the alternate bar from the edge of the wing is cut so as to form the edges with fly meshes. After making 19 knots at the base, Babylon is fabricated as shown in Fig. 61.

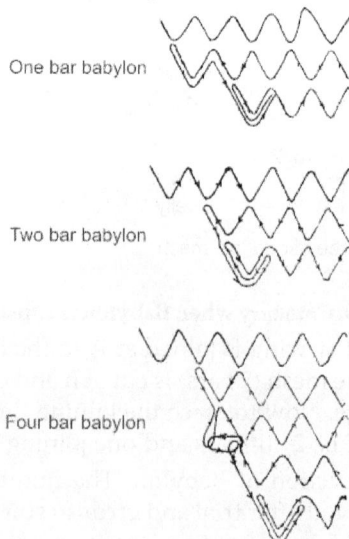

One bar babylon

Two bar babylon

Four bar babylon

Fig. 61: Construction of Babylon

At the end of Babylon, 2 rows of meshes at the edge (D) are fabricated at a rate of 13 or 14 meshes which help to reduce the distortion of wing meshes hung after stretching (outstretch). All these fabrications are done only with double meshes. At the end of wing one bar is made double.

Hanging: At bosom near Babylon, five meshes are hung closer than normal hanging which reduce the strain at the quarter (Fig. 62). The hanging looseness Babylon is about a bar length in each section, i.e., in one bar, 2 bar and 4 bar Babylon. Although the meshes in the wing end after Babylon are stretched, it does not effect the meshes below one row because of the fabrication of 13 double meshes on 14 meshes. The length outstretch 2.25 m in the design means the difference in length of wing vertical measurement (8m) and diagonal measurement (b and c) after stretching which is 10.25. Practically it was found successful in hanging the meshes after outstretch. The total hanging length of 10.25 m is actually hung in 10.10 m, the little difference here is caused by a bar looseness given in the Babylon section. Hanging is directly done on framing lines, except in one foot rope where bolch line is used.

The horizontal meshes at the end of wings are not hung, but are laced together with the leg. The upper or lower and side hanging ropes are tied at the end of wing to form the leg.

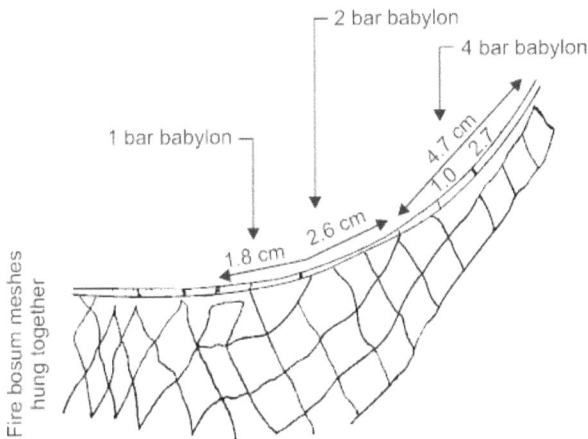

Fig. 62 : Babylon hanging for 6 x 6F mid water trawl

Rectangular mid water trawl : A notable difference in the design of rectangular trawl is that the lower panel is little longer than the upper panel and therefore the side panel wings are asymmetrical. Scharfe (1964) notes that the rectangular shape was chosen because it was difficult to open sufficiently in vertical direction a net with equal panels. This forward extension of bottom panel is meant to give the net a downward sheering tendency to secure good bottom contact, if desired. The performance of rectangular net was found better when the outstretch of the bottom panel is increased.

Rigging : A stern trawler equipped with bottom trawling can also do mid water trawling. The additional equipments for mid water trawling are mid water trawl, otter doors, and trawl sonde. The trawl sonde is essential to observe the action and performance of the gear, as well as the behavior and distribution of fish around the net opening. By using echosounder of the vessel and a chart prepared from the warp angle measured by a spirit level reading on a protractor, the position of the shoal and depth of the trawl in operation can approximately be known.

The depth of the traw! is calculated trigonometrically Sin equals opposite side by hypotenuse. If the warp released is 75 fathom from sea surface and the angle measured is 15 degree, then depth of trawl will be (Fig. 63)

Sin 15 degree = BC/AB; Sin 15 degree = 0.2586 x 75 = 19.5 fathom Fig. 63

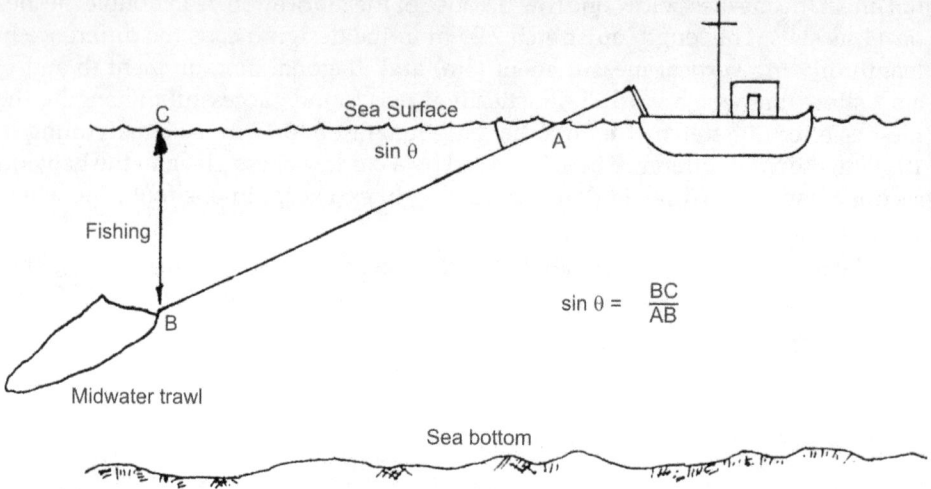

Fig. 63: Calculation of fishing depth of trawl

Two types of otter doors are used for one boat mid water trawling- the suberkrub door and 4 door mid water boards. The rectangular mid water trawl is rigged with suberkrub otter boards and square (quadratic) trawl is assembled with four door mid water boards. In the 4 door type, the two top boards are smaller and lighter than the two bottom boards. On the top edge of the top board a polyform float is attached to increase the buoyancy and stability. From a 200 HP boat a 6 x 6 fathom mid water trawl can be rigged with a pair of 90 x 53 cm upper door (18 kg each) and a pair of 170 x 90 cm lower door (128 kg each). The rigging details is shown in Fig. 64.

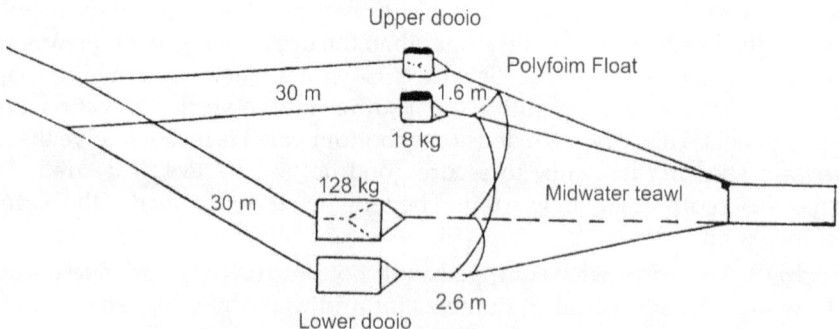

Fig. 64: Rigging of single boat mid water trawl

Unlike bottom trawl, mid water trawls are opened downward because of the weight attached between leg and bridle. Since the pulling force of the gear act on the head line, static buoyancy can not contribute much to the vertical opening of the trawl. So the floats attached to the head line are to keep the net clear during shooting and hauling. Since the meshes in the front part of mid water trawls are big, there are chances to foul the webbing with floats and so floats are arranged in dense pack covered with small netting and attached to the head line.

One boat mid water trawling trials conducted in south west coast of India

The trawl used for the operation was a Norwegian made 12 x 12 fathom mid water trawl (four equal panel). The inside of the cod end was laced with a cover having 8 mm stretched mesh size. The design details of the net are given in Fig. 65.

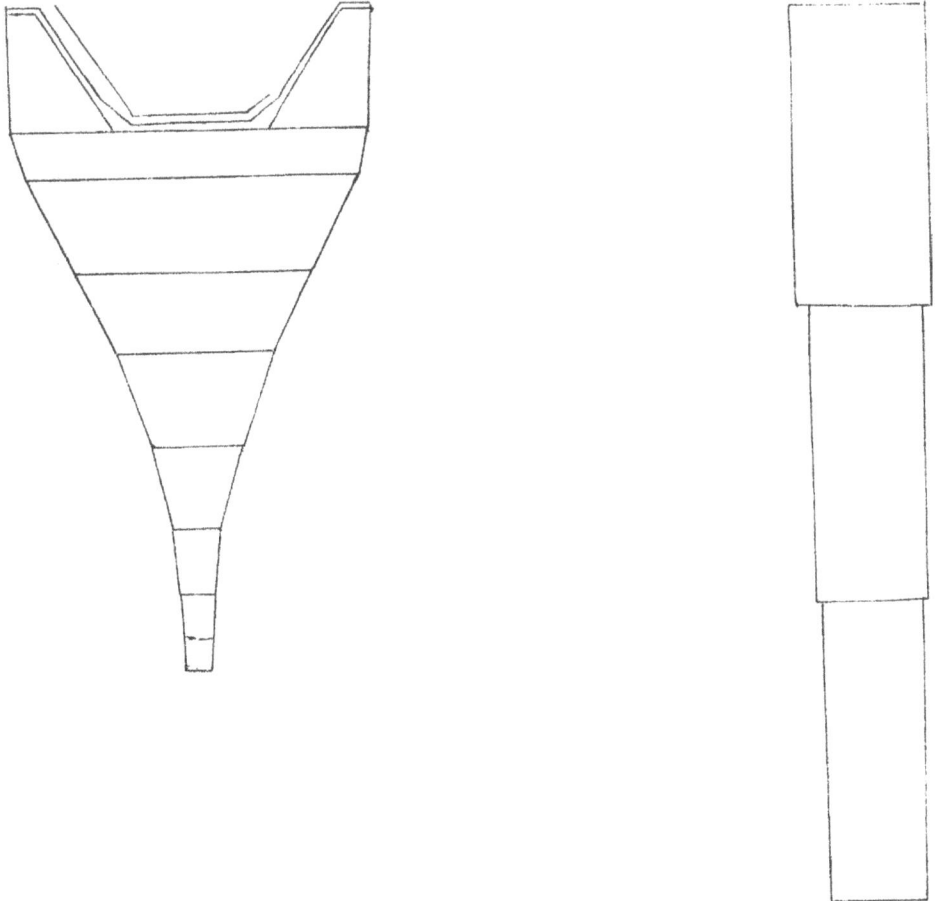

Fig. 65: Experimental mid water trawl

On the head line of the net, a SIMRAD trawl eye 50 KHz was attached (Fig. 66).

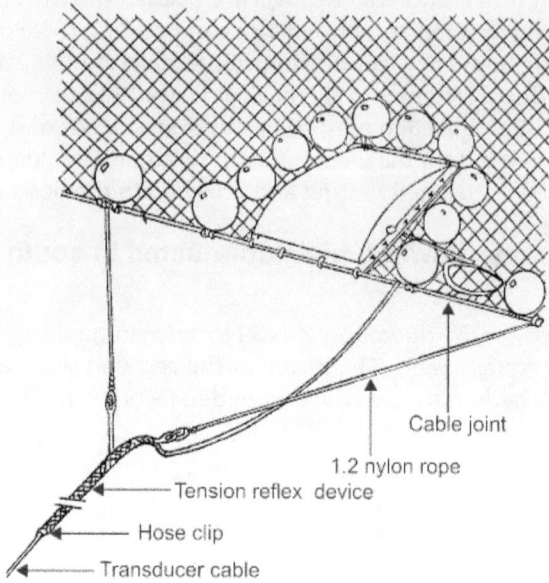

Fig. 66: Fastening the transducer to the trawl head line.

The recording unit (SIMRAD EXC) was attached to the transducer by cable having a length of 300 m. A 36 KHz Skipper Sounder was also used for observing and comparing the fish school with the trawl sonde. The net was rigged with four rectangular flat doors (1.4 x 2.8 m and 1.0 x 1.6 m respectively) and 40 m sweep wire. The details of trawl door are given in Fig. 67.

16 mm dia through bolt
50 × 6 mm M.S. Flat 50 × 15 mm MS. Flats 4 Nas
280 cm
140 cm
5
Anlili wood planks 150 80 mm castiron
Weight of one door 363 kg.
8

155 cm 220 cm 125 cm
½" dia G.I. chain
155 cm 225 cm 140 cm

16 mm dia bolt
162
102
4
Anjili wool plank 48 × 6 mm m.s. Flat alround
Weight of one door 55 kg

125 cm 75 cm 75 cm
125 cm 115 cm 75 cm
½" dia G.I. chain

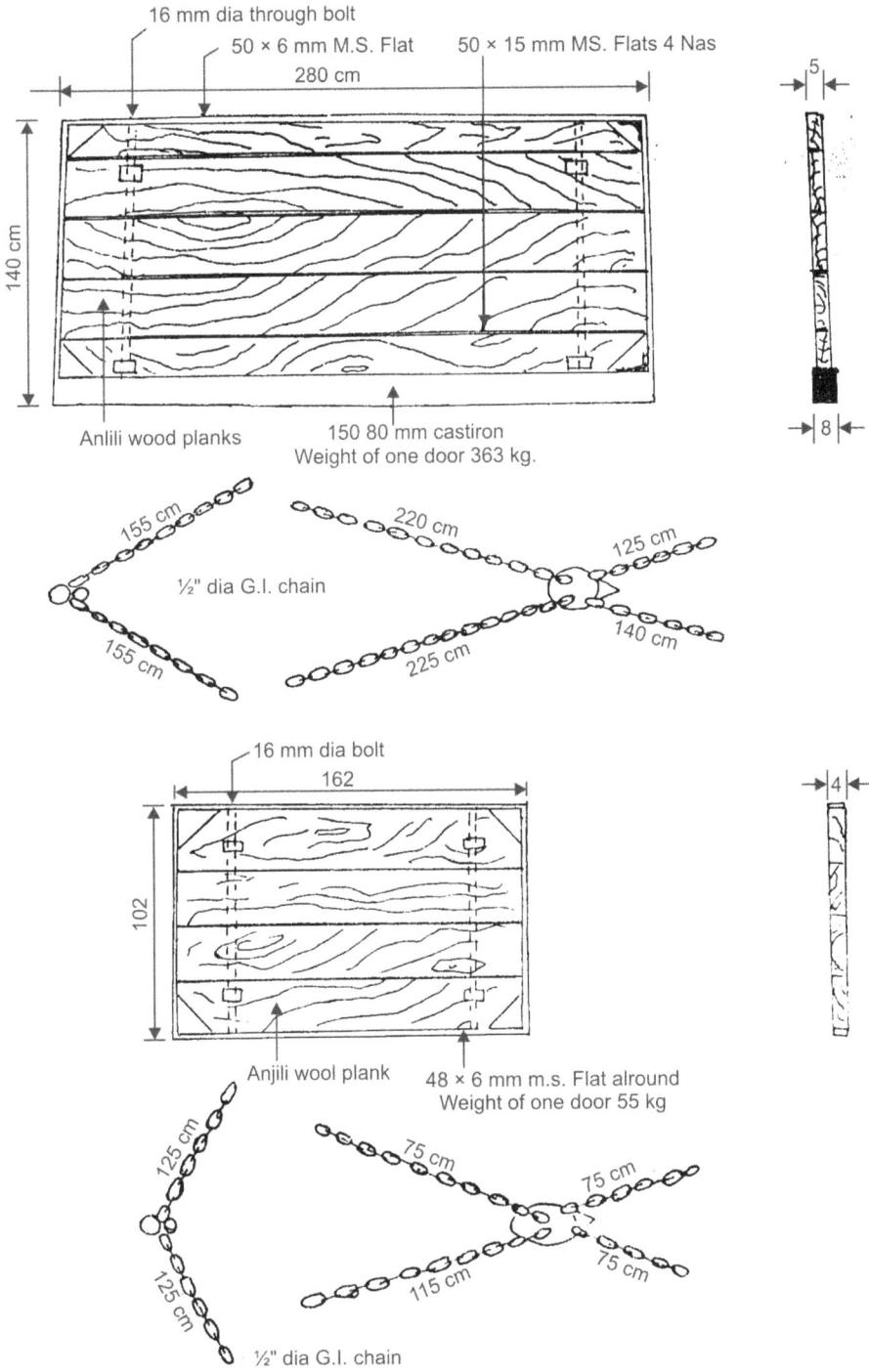

Fig. 67: Mid water otter doors

The rigging details of the gear are diagrammatically presented in Fig. 68.

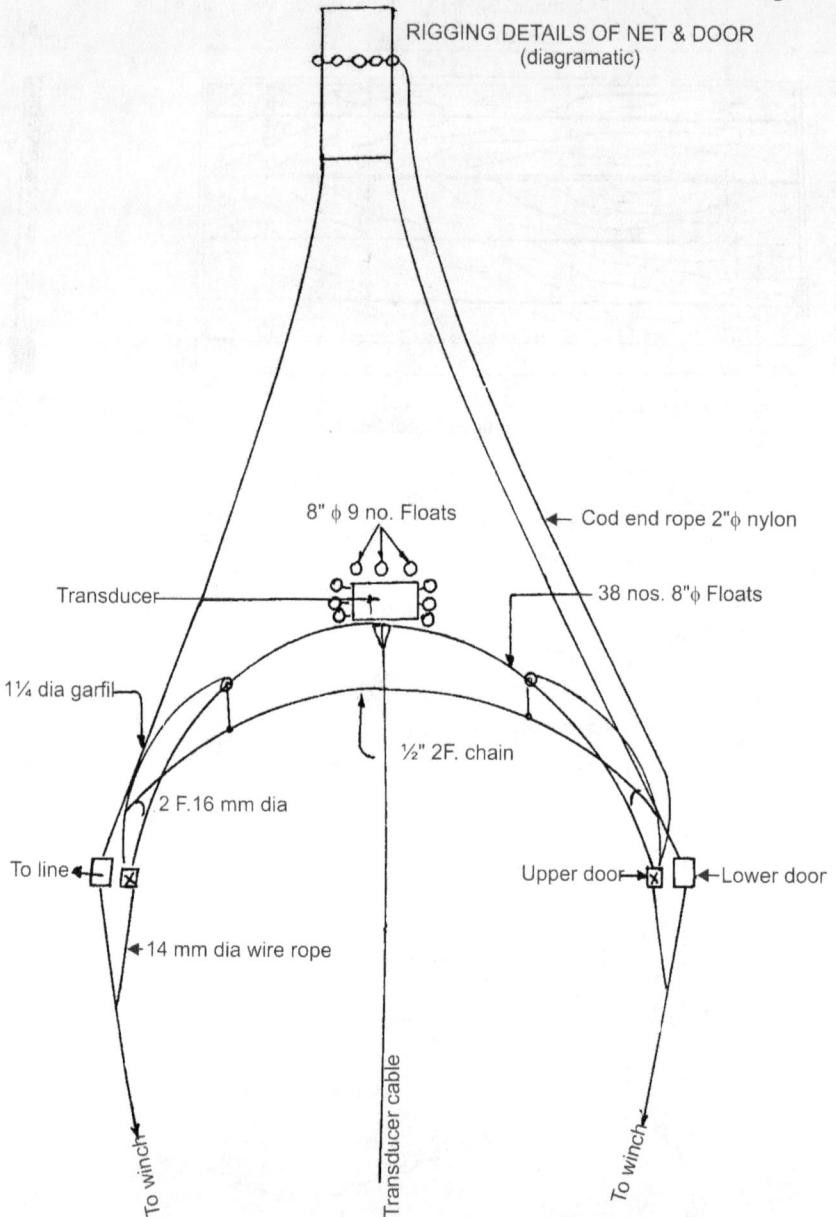

RIGGING DETAILS OF NET & DOOR
(diagramatic)

8" φ 9 no. Floats

Cod end rope 2"φ nylon

Transducer

38 nos. 8"φ Floats

1¼ dia garfil

½" 2F. chain

2 F.16 mm dia

To line

Upper door

Lower door

14 mm dia wire rope

To winch

Transducer cable

To winch

Fig. 68: Rigging details of net and doors.

The operations were done in the depth range of 16 to 22 fathoms, when about 3 times warp was paid. The net was not completely opening as was seen from the recording unit. Therefore warp was released to 50 to 75 fathoms so as to cover almost the entire column of the depth. The cable of trawl sonde was provisionally operated by hand.

The trial showed that the trawl could not well be lifted off bottom with the towing power available with the vessel (600 BHP). Consequently trial was restricted to the high opening bottom trawling with an opening height from 14 to 20 m. The front opening was presumed to be in the form of a trapezium having the horizontal distance more in the bottom than in the head line. This is mainly due to the smaller boards rigged on the head line and the heavy larger board in the lower rope.

The result indicate that single boat mid water trawling may prove to be viable fishing technique for Anchoviella.

One boat mid water trawl

To find out a suitable gear for pelagic exploitable stocks (Sardines, Mackerels and Anchoviella), the Norwegian trawl (19 x 12 fathom mid water trawl with four equal panels) was used. The inside of cod end was laced with a cover having 8 mm stretched mesh size. On the head line of the net a SIMRAD trawl eye 50 KHz was attached. The recording unit (SIMRAD EXC) was attached to the transducer by cable having a length of 400 m. A 36 KHz Skipper echo-sounder was used for observing and comparing the fish school with the trawl sonde. The net was rigged with four rectangular flat doors (1.4 x 2.8 m and 1 x 1.6 m respectively) and 40 m sweep wire. Fishing operations were conducted from a 93 feet vessel powered with 600 BHP engine. Trial fishing was conducted in the depth range of 16 to 22 fathom. Three times warp was paid, but the net was not completely open. The warp was further released so as to cover almost the entire column of the depth.

Details of operation of 19 x 12 fathom mid water trawl.

Cruise	Haul no.	1	2	3	4
1.	Position	7-77/6D	8-77/1E	8-77/1E	7-77/6C
	Depth	16F	16 F	17F	22 F
	Warp(F)	50	75	75	75
	Time	1 .24 hours	2hours 10min.	1 hour	1.5 hours
	Direction	90 degree	270 degree	90 degree	90 degree
	Speed (knot)	2.5	2.5	2.5	2.5
	Catch	Nil	Nil	0.4 tonne	3.0 tonne
		Total catch - 3.4 tonnes			
2.	Position	7-77/6D	7-77/6D	7-77/6D	7-77/6D	7-77/6D
	Depth	18 F	18F	17F	17 F	17 F
	Warp(F)	75	75	75	75	75
	Time	1 hour	1.5 hour	1.5 hour	2.5 hour	45 min.
	Direction	90 degree	90 degree	90 degree	90 degree	90 degree
	Speed (Knots)	2.5	2.5	2.5	2.5	2.5
	Catch	9.7 tonnes	nil	nil	nil	nil

The cod end was full of jelly fish which was thrown overboard.

A catch of 3.4 tonnes during two hauls of 2 hours time (haul 3 and 4) was obtained in the first cruise. During the second cruise, the vessel caught 9.7 tonnes in the first haul itself. The echoes were fairly good and a strong recording of *Anchoviella* and sardines were found. The trawl could not well be lifted off bottom with the available towing power of the vessel. So the trial was restricted to the high opening bottom trawling with opening height from 14 to 20 m. The front opening was presumed to be in the form of a trapezium having the horizontal distance more in the bottom than in the head line. This is mainly due to the smaller boards rigged on the head line and the heavy larger boards on the lower rope.

10.5 m mid water trawl: A 10.5 m nylon, four seam, equal paneled trawl net and Japanese vertical curved otter boards were engaged for carrying out fishing operation. Particulars of the accessory gear used during the course of these trials are;

Head rope – Manila, 10.5 m, Dia. 12.7 mm

Foot rope–Manila, 10.5 m, Dia. 12.7mm

Sweeps Head rope – Manila, 12 m , Dia. 12.7 mm

Sweeps Foot rope – Manila, 12m,Dia. 12.7mm

Floats-9 spherical, aluminium alloy, Dia– 10.16 cm.

Total extra buoyancy along the head rope – 3.6 kg

Total weight along the foot rope – 6 kg (570 g/m)

Otter boards – Vertical curved, size - 100 cm x 50 cm, thickness 20 mm, weight in air 25 kg each (Fig. 69 and Fig. 70)

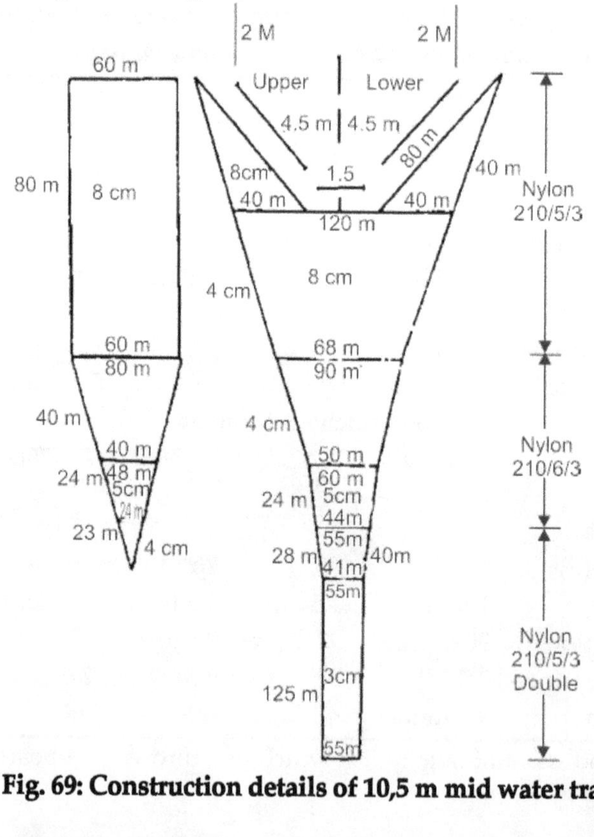

Fig. 69: Construction details of 10,5 m mid water trawl

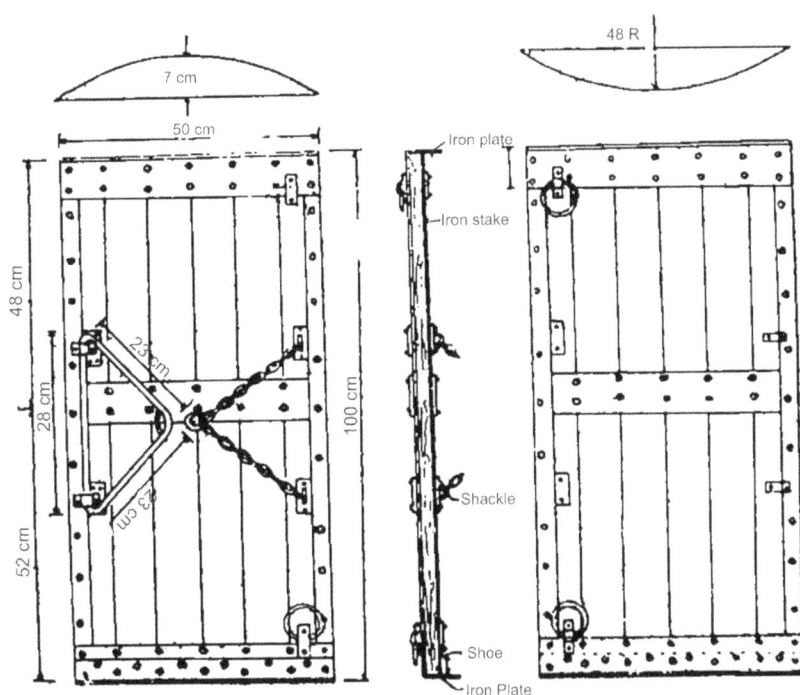

Fig. 70: Design details of vertical curved otter boards

In nine days, 50 hours of trawling, 2940 kg fish and 9 kg of prawn was landed by 10.5 m mid water trawl from a depth of 23 to 39 m. Sciaenids constituted about 38% of the total catch followed by Indian salmon (21.7%) and Lactarius (16.3%).

The average horizontal distance between two otter boards in action was 17.15m which worked out to 50% of the head rope length including sweeps. The towing resistance offered by the entire gear while in operation was 476 kg.

Two boat mid water trawl (Pair trawling)

Two boat mid water trawling was tried in south west coast of India as an effort for diversification of fishing program, mainly for pelagic resources during 1973. Two 32 footers of 9.76 m (LOA) and 2.9 m beam having 48 BHP engine with 7 tons displacement were selected for the purpose.

Both the boats were having the same type of gallows and winches. The starboard winch drum of one vessel and the port side winch drum of the other carry on each 220 m of 8 mm dia, 6x19 construction wire rope. The wire rope passes through the gallows which are on the same side of the winch drum.

Fishing gear: The net designed for this trial is a four panel rectangular net made of polyamide and colorless webbing. The length of the head rope of the net was 58 feet (12.6 m) having 758 meshes of 132 mm mesh size around bellies. In order to catch Anchoviella resources, a mesh size of 8 mm stretched length has been selected for the cod end region. The details are given in Fig.71.

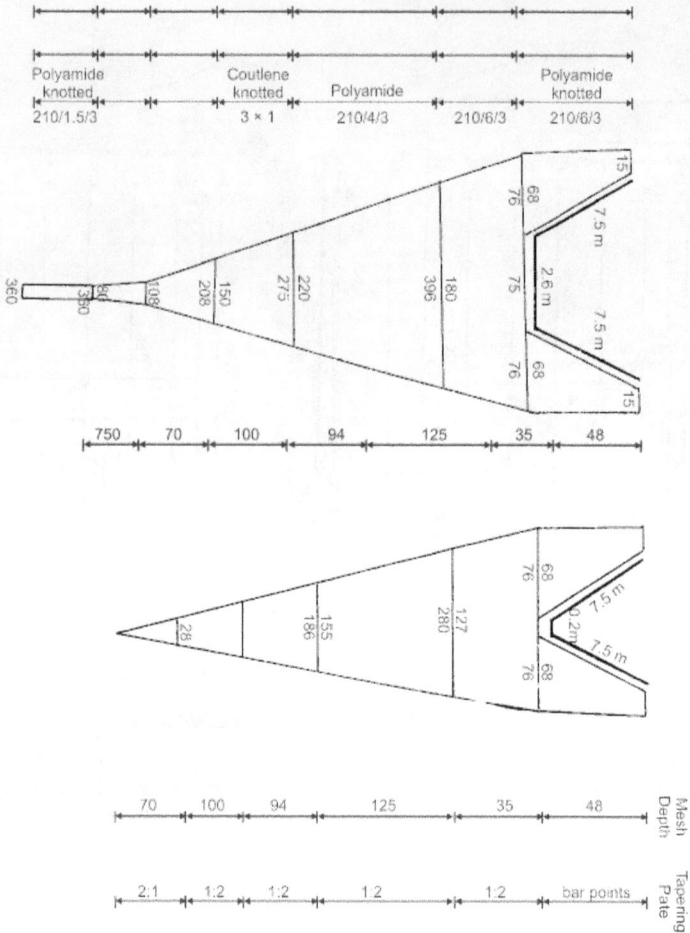

Fig. 71 Two boat pelagic trawl for 32 feet boat (drawing not to scale)

In constructing the trawl, the bosom corners of each panel is closely mounted with 10 meshes and the bosom corner edge piece is made of double twine to distribute the strain. Three double meshes are provided as selvedges on the mounting side all over 9 PVC floats of 200 mm diameter are attached to the head line at distances of 50 cm, I m, 1.5 m and 3.3 m from bosom center to wing tip. The foot rope is attached with 17 kg m of lead and chain arranged at equal intervals. On the mounting rope of each of the side panel, 3 kg of chain is mounted in two pieces. The cod end halving becket is connected to the head rope wing tip with 37 mm circumference polyethylene rope. G. hook and link are attached on the legs for easy transfer of wing from one boat to the other.

Rigging : Thirty meter of sweep (31 mm circumference, manila combination rope) is connected between the head rope legs and warps. The foot rope sweep is of 37 mm circumference manila combination rope and in two pieces of 7 and 24 m. The 7 m rope is connected to the foot rope leg at one end and an iron weight of 20 kg at the other end (Fig. 72). The 24 m rope is connected between the weight and the warp. Both head rope and foot rope sweep lines are joined to the warp with swivels.

The net and one set of sweep line both head side and bottom side together with weights are rigged on the towing vessel. The other set of sweep line and weight with connection to the warp are carried in the other vessel. The details of the attachment in the position of weight ©, detachment of net from one vessel to the other (A) and (B) are presented in the figure 72.

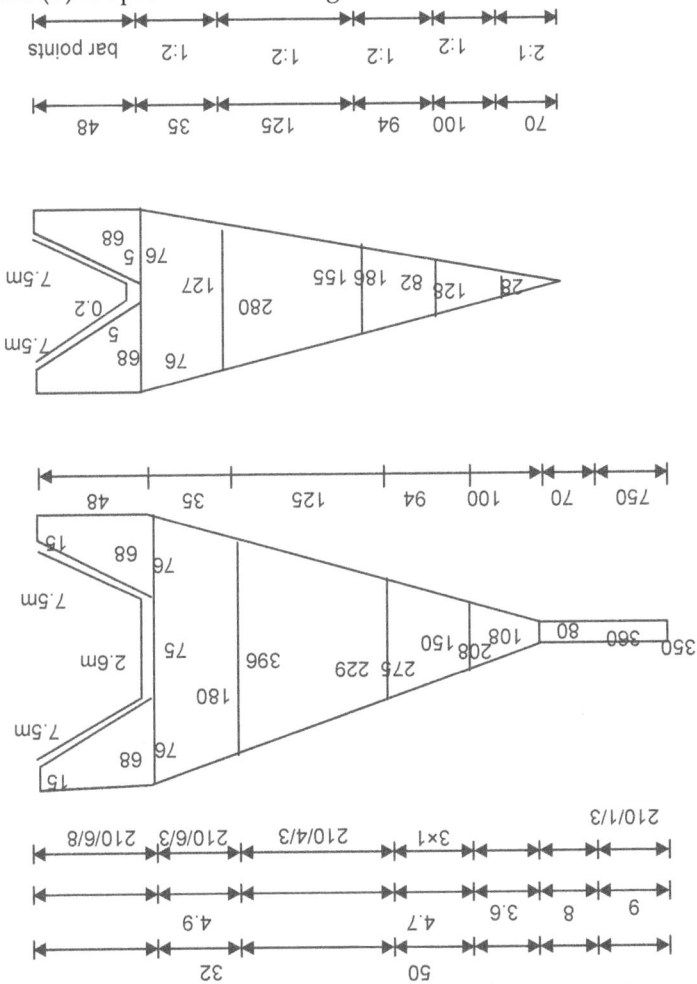

Fig. 72: Rigging of two boat mid water trawl

Operation: The towing vessel chooses the fising ground and course and directs the operation and heaves the net. As soon as the fishing area is selected, the towing vessel pays out the net from cod end to wing. Both set of wing ends are tied on the towing vessel. The second vessel approaches the main vessel from the side and picks up one set of wing end, i.e., head and foot side wing ends and connects the same to their respective lines. Both the vessel set up speed gradually and move straight away from each other and after paying the sweep line and the predetermined length of warps, they begin to move in parallel courses. In the initial stages 60 m rope was usually tied between the two bows in order to keep the two vessels at equal distance. As the operation progressed, the crew could maintain the distance between the two boats even without this rope. Therefore in the second year of operation the rope at the bow was given up. During the operation, course, speed and distance of vessels are maintained constant.

After the desired time, the two vessels slow their engines and heave in the warps and sweep lines lying in the same course. At the end both the vessels move towards each other. The second vessel hand over the previously received legs of head and foot rope of the net after releasing from the sweep line. The net is then hauled in the towing vessel.

Fishes during southwest monsoon period are concentrated in the off bottom and surface layers due to strong upwelling of oxygen deficient water on the shelf of south west coast, the small crafts, which depend on bottom trawling are laid up due to poor fishing results. Mid water pair trawling may be found to be a promising technique to the small crafts.

During the operation, the horizontal opening of the net was calculated from the distance between the boats, the length of net and the length of warp.

At a distance of 40 m, between the boats, the horizontal spread of the net was 12.08 m which is 68.3% of the horizontal spread as worked out below;

$$\text{Horizontal opening} - \frac{40 \times 30.1}{(30.1+36.6+33)} = \frac{1204}{99.9} = 12.08 = 68.3\%$$

Where, 30.1 m = length of net, 36.6m = warp length and

33 m = sweep length and leg

It is therefore assesed from the catch and spread of the net that 40% of towing line and length of net may be suitable as the distance between the two boats.

The average catch per hour was only 39.5 kg (not an appreciable quantity). This was mainly due to the difficulty of maneuvering the vessel as the crew were new to the gear and technique of operation. When the crew became experienced, the operation was perfected and in 1974 (next year) operations gave an average catch of 96.9 kg per hour. During the period of June to October, 1974 the yield went up to a maximum of 3.2 tonnes per day. Of the total catch, about 53% was shallow water mix, 16% was sardine, the remaining quantities were shark, ray, prawn and cuttle fish.

6 | Bottom Trawl
Large Meshed Bottom Trawl

Trawling has become an established fishing method in India over the period since its introduction in 1960s. There are about 35230 trawlers operated in the country of various sizes ranging from 8.5 to 23.2 m LOA with engines ranging from 90 to 420 HP at the rate of 2000 rpm. Bottom trawling on a commercial level started at Cochin during early 1960s and it gradually spread to other coastal states. The Central Institute of Fisheries Technology (GIFT) played a major role in introduction and popularization of various trawls into fishing industry of the country. There has been a significant progress in the design and performance of trawl system since its introduction.. The mesh size, particularly in the fore parts of the trawl net has undergone significant changes in the course of time along with increase in dimensions of the net. Over the years GIFT has designed several trawls, such as, two-seam trawl, four-seam trawl, six seam trawl, long winged trawl, bulged belly trawl, high opening trawl and semi-pelagic trawl and introduced energy saving concepts in trawl design, such as, large mesh trawl and rope trawl for bottom trawling. Large mesh trawls were fully adopted by the fishermen and there have been a significant increase in the mesh sizes in the fore parts of the trawl net owing to its energy saving benefits and efficiency.

The advantages of using large meshes in the fore part of the trawl net are (a) increased flow and quick filtration of water, (b) reduction in drag time during tow, low fuel consumption, ease of fabrication and maintenance and (c) reduction in total weight of netting and saving in material cost as compared to the conventional trawls of the same size.

A trawl net is basically a conical shaped gear towed through water, the mouth of which is kept open horizontally by means of a beam or otter boards and vertically by means of floats, kites and sinkers. In the case of pair trawling, the horizontal mouth opening is maintained by two vessels operating at an approximate distance from each other.

Since its introduction, the trawl net has undergone many modifications in its design, fabrication and operation depending on the type of fishes caught, local conditions, size and engine power of the trawler employed for fishing. The size of the trawl net has increased considerably along with the size of trawlers.

Factors, like mesh size, shaping of the net, assembling and mounting play an important role in determining the efficiency of the fishing gear. A good design and fabrication skills are essential pre-requisites for producing a good trawl net. The net makers have standardized a few designs which have been successful in the past and well accepted by the fishermen. But net makers usually keep on modifying the original designs in an attempt to improve performance to catch target fishes. Usually the length of head rope is increased to enhance the trawl mouth opening.

The introduction of large meshes in the fore parts has further increased the possibility for increasing the length of head rope without increasing the drag.

After the introduction of synthetic fibers in the fishing industry in the early 1960s, all the trawl nets are fabricated exclusively using high density polyethylene (HDPE) twines of different twine sizes. The earlier practice of fabrication of trawl net by hand braiding has completely ceased as ready-made netting has become readily available in the market. This is a boon to the fishermen as it saves lots of time and labor. The ready-made nettings are then tailored to the panels of required shape, assembled and rigged to produce trawl net. The cod-end is constructed using either hand-braided or machine made netting of mesh size ranging from 12 to 15 mm.

Each vessel on an average carries about 15 trawl nets of various designs for harvesting shrimps, cephalopods and fishes. The average life of each trawl net is about 3 years and every year a large number of trawl nets are required for replacement of the used ones.

Mesh sizes of large mesh bottom trawls in the country.

Year	Head rope (m)	Mesh size in mm			
		Wing	Square	Body	Cod end
1979	32	150	150	150-40	30*
1988	25	200	200	160-30	10-15
1989	25	200	200	160-40	30*
1994	54	400	400	400-25	10-15
1996	33.7	400	200	200-40	40*
1998	40	200	150	150-40	30*
1998	32	200	200	150-40	30*
1998	20	300	300	200-50	30*
2004	35	400	400	200-40	10-15
2007	45	800	1000	1000-40	10-15
2010	45	1000	1000	800-40	10-15
2010	50	1500	1500	1000-40	10-15
2011	59	5000	5000	1500-40	10-15

There is a tendency among fishermen to give fancy names to different designs, such as, multi-purpose trawl net, Bombay net, Koti net, Laden net, Saddam net, Cuttle fish net, Shrimp net, Red ring net, Rani net, Lazard fish net, Ribbon fish net, Japan net (4 seam), Disco net (high opening trawl) Speed trawl (more than 1000 mm mesh size in front sections) etc.

The large mesh trawl net concept was introduced by CIFT in the late 1970s with 150 mm mesh size in the wing portion of the trawl net. Use of ropes in foreparts of the net for herding the fish was also tried upon and the fuel consumption was reported to be less. The concept of large mesh was accepted by the fishermen only during late 1980s and it gained popularity and spread throughout the coastal states.

Trawl nets can be selective in catching certain fast swimming species of fish based on the design and rigging pattern. It was found that large mesh trawl is efficient in catching quality fishes in Gujrat State. Mesh size of netting used in the front panel sections have increased up to 5000 mm to increase the trawl mouth area without increasing the drag or alternatively to increase the towing speed in some designs of fish trawl.

In late 1990s the fishermen of Gujrat started using mesh size of 400 mm in the fore parts of the trawl. During that period large mesh netting were not readily available in the market and the fishermen had to fabricate large mesh netting by hand. With increasing demand for large mesh webbings, ready-made large mesh netting entered the market. With the use of large mesh sizes there was a proportionate increase in the head rope length of trawls.

Over the years, trawls with large mesh sizes in the fore parts of the net became popular along the Indian coast. With the increased size of fishing vessel and increased installed engine power, the mesh size of the fore part of the trawls increased substantially and presently mesh sizes of 5000 mm are being used for high speed trawling targeting fast moving off-bottom fishes- Mesh sizes ranging from 4000 to 5000 mm are being used in trawls operated from Goa, Karnataka and Kerala.

The trawling speed of conventional trawl is 2 to 2.5 knots. However, with the advent of high powered engines and large mesh trawls, the trawlers drag their trawling gear at high speeds ranging from 4.5 to 5 knots, targeting fast moving fishes. But though there is an increase in mesh size in fore parts of the trawl net, the cod end mesh sizes have remained the same and even have been reduced to 10 mm in some cases. The use of small mesh size in the cod end of the trawl net are preferred by the fishermen as the by catch and trash fishes generated are in demand for fish meal production and manure. But this tendency has severe ecological consequences as the by-catch also includes significant quantities of juveniles of commercially important fishes which will negatively impact on the commercial fisheries.

Traditional hand braiding of trawl nets has completely ceased and this has been replaced by ready made netting. The net making industry is able to cater to the varied mesh size requirements of the fishermen. With the increasing number of trawlers in the country, there is a great demand for good designs of ready made trawl nets. Many net makers have established themselves in the profession over the years and have gained experience in the art of designing and fabricating trawl nets by keeping the wastage of netting materials to the minimum. Many net makers have ready made nets of various designs and sizes appropriate for the size and installed engine power of the vessel. They keep on modifying and refining the design depending on the feed back they get from fishermen, but important factors like the drag, suitability to the installed engine power and the size of the vessel are not considered while these modifications are made.

Choosing the right size of trawl based on the power of the vessel, relationship between mesh size and twine size for trawls is extremely important for efficient performance of the trawl gear .The size, type, weight and rigging of otter boards should be appropriate for the trawl net and the horse power of the vessel. These are not generally followed by the fishermen and they rely only on trial and error methods which are most often not successful and the net makers keep on changing the design of the trawl net. The fishermen are always inclined to have bigger boats and higher horse power to operate increasingly larger nets which may not always be appropriate for economic fishing operation. A study on the right sizing the different components of the fishing system for economical and sustainable operation has to be carried out. Mesh size in the cod end has to be regulated and the use of stipulated cod end mesh sizes has to be enforced for responsible fishing.

The concept of large mesh trawl will lead to energy saving in trawling provided standardization in fishing systems in terms of capacities takes place.

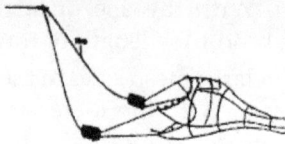

Fig. 73: A demersal trawl net

Design criteria of bottom trawls for 16.15 m steel trawler powered by 170 HP engine

The Poland built 16.15 m steel trawler with 170 HP diesel engine was developing a cruising speed of 8 knots at 1800 rpm. For side trawling two gallows were fixed, one at fore and the other at aft of starboard side of the vessel.

The conventional trawl nets used from these vessels in combination with single slitted oval type otter boards of 146 x 86 cm, weighing 80-86 kg were 24.4 m (89 feet) two seam and 22.3 m (73 feet) and 25.6 m (84 feet) four seam trawl net. All these nets were having short belly and throat pieces. The total length of the net from mouth to cod end was 50-55% of the head rope length. The width of the side panel of four seam net was very small and not proportionate to the width of upper and lower panels.

These gears and accessories operated from 16.15 m steel trawlers were found to be under-sized in relation to the power developed by the engine as the rpm indicated during trawling was only 800 to 900 at a trawling speed of 2 to 2.5 knots which is 45 -50% of the total rpm of the engine. This necessitate the developing of larger and standard size gear for utilizing the maximum power available.

For optimum power utilization 32 m long wing and bulged belly trawls were designed and developed for the capture of bottom and off-bottom fishes including shrimps. The design, construction details and method of rigging are given in Fig. 73 and Fig. 74.

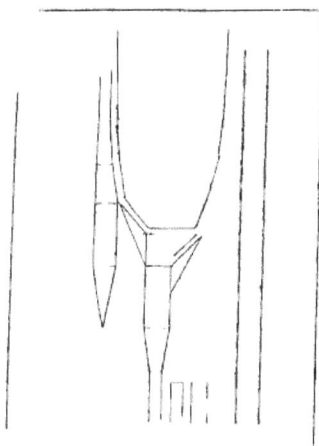

Fig. 74: 32 m long wing trawl for Polish trawler (170 HP engine)

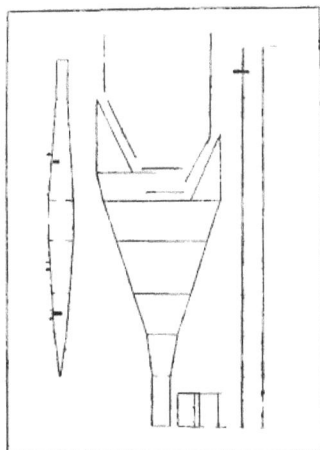

Fig. 75: Bulged belly trawl for Polish trawler (170 HP engine).

The details of total number of meshes prepared and the labor involved for the fabrication, assembling and mounting of these trawls are shown in the following table.

No.	Particulars of work	Long wing	Bulged belly
1.	Head rope length of trawl net (m)	32	32
2.	No. of meshes fabricated (single knot)	401600	1026400
3.	Persons engaged/day for fabrication	44	44
4.	Days required for fabrication	6	14
5.	Persons engaged/day for assembling & mounting	5	5
6.	Days taken for assembling and mounting work	3	6

The otter doors used were heavier, pairs of rectangular type of size 182 x 91 cm weighing 130kg.

Particulars of accessories required for rigging trawl gear of each category are;

No.	Name of parts	Long wing trawl 32m	Bulged belly trawl 32m
1.	Head rope, material, size, length	Manila, 25 mm, 40m	Manila, 25 mm, 40 m
2.	Foot rope, material, size, length	Manila, 25mm, 45 m	Manila, 25mm, 45 m
3.	Bolch line, material, size, length	Monofilament, 6 mm, 80 m	Monofilament, 6 mm, 80 m
4.	Side, rope, material, size length	Monofilament, 10 mm 60 m	Monofilament, 10mm 130 m
5.	Floats, material, shape, dia. Extra bouyancy, no used	Aluminum, round 152 mm 1600g, 15 nos	Aluminum, round 1600 g, 15 nos
6.	Sinkers, material, shape, size,	Lead, spindle, 30 mm dia hole, 400 g each	Lead, spindle, 30 mm dia hole, 400 g each
7.	Total no used	87	73
8.	Weight in kg	34.8	29.2

Matching power in relation to the size of trawling gear

Normally during trawling the engine rpm can be kept at 75% of the total rated rpm for a proper loading of the engine. An observation of their rpm developed (800-900 at a trawling speed of 2-2.5 knots) during towing the existing gear system showed that the vessel was rigged with under sized gears and accessories. When the same trawler was rigged with larger gear and accessories (32 m long wing type and 32 m bulged belly type), the power utilized was increased to proper loading of the engine as indicated by the increase in rpm from 800 -900 to 1200-1300 at a trawling speed of 2-2.5 knots. The increase in catch (from I 395 kg/day to 2525 kg/day on an average) substantiates the fact that the gear and accessories developed was suitable and of standard size for the trawler of 16.15 m steel trawler.

Exploratory bottom trawl trials off Tamil Nadu in 1969

A total of 1100 square miles area was sampled off Tamil Nadu in depths up to 30 fathoms by the trawling gear during the months of February, April, May and July for 89 days with a fishing effort of 365 hours. The total catch accounted for 95560 tonnes, which works out to 262 kg per hour of fishing effort.

The details of the steel trawler which was engaged in the exploratory fishing program are given below.

Length OAL (m) – 22.5; Breadth (m)–5.5; Depth (m)–3.3, Tonnage (gross)–69.15; Speed – 10 knots; Crew – 10, Fish hold (tons) – 15; Main engine – 262 HP at 900 rpm; Auxiliary engine – 16 HP; Trawl winch – Mechanical two drum winch, hauling speed 40 m per minute; Variable pitch propeller.

Sisal trawl of 15 and 24 meter head rope length as well as monofilament trawl of 35 m head rope were experimented off Tamil Nadu from a 22.5 m vessel of 69.15 gross ton. The vessel was powered with 262 HP main engine developing a speed of 10 knots at 900 rpm. The vessel had a Kelvin Hughes-Cores fish finder and a variable pitch propeller.

The details of these trawl nets and riggings are furnished below.

15 m trawl was made up of 2 mm dia. Sisal twine, having a mesh size of 40-70 mm bar. 8 mm dia. wire rope served with sisal twine of 2 mm was used as head rope which was 16 m in length. Foot rope of 23.6 m was made up of 9 mm dia. wire rope with 2 mm dia. sisal twine. Bolch line was made up of one and one fourth inch circumference manila rope. Floats of aluminum alloy having 6 and 8 inch diameter was used as floats with a total buoyancy of 30 kg. 11 mm dia. wire rope wound with 51 mm dia. Manila rope with wooden rollers served as ground rope. Hanging of net webbings were 50% for bosom and 12% for upper and lower wing. The weight of netting was 75 kg. 1.3 square meter was the area of the otter board weighing 150 kg provided with 30 m sweep line of 9 mm diameter. The design details of 15 meter trawl are given in Fig. 76.

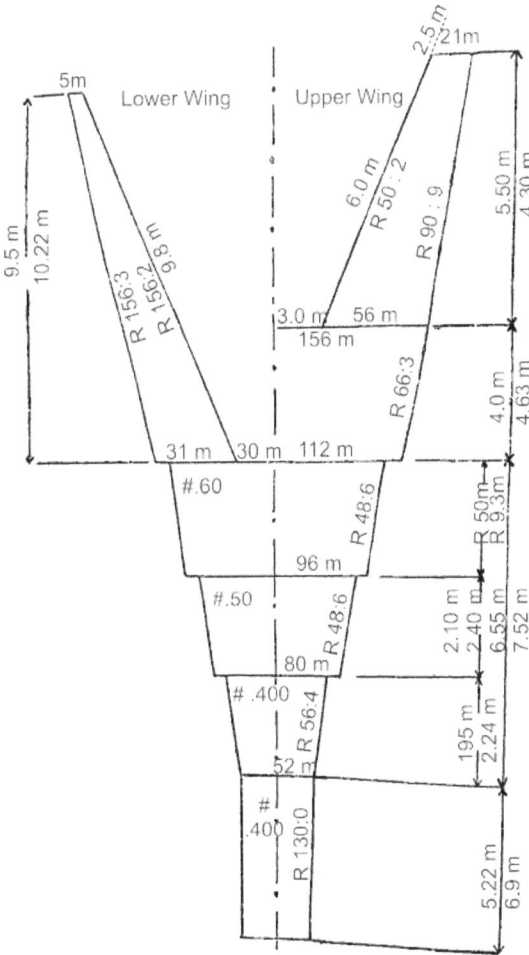

Fig. 76: Design diagram of 15 meter bottom trawl

Twenty four meter trawl was also made up with 2 mm dia. sisal twine having a weight of 105 kg. The mesh size was 25-70 mm bar. 8 mm dia. wire rope wound with sisal twine of 25 m length served as head rope. 33 m long 9 mm dia. wire rope served with sisal twine was provided as foot rope. The bolch line was made up of 6 mm dia. monofilament rope. Floats of 6 and 8 inches dia., made of aluminum alloy of 45 kg total buoyancy was provided in the head rope. 11 mm dia. wire rope wound with 51 mm dia Manila rope and with 4 inch dia. rubber discs served as ground rope. Fifty percent hanging was provided at bosom, while 12% hanging was given both for upper and lower wings. The otter boards were 2 square meter having a weight of 200 kg. Sweep line was of 9 mm dia. and 40 meters in length. The construction details of 24 m bottom trawl is given in Fig. 77.

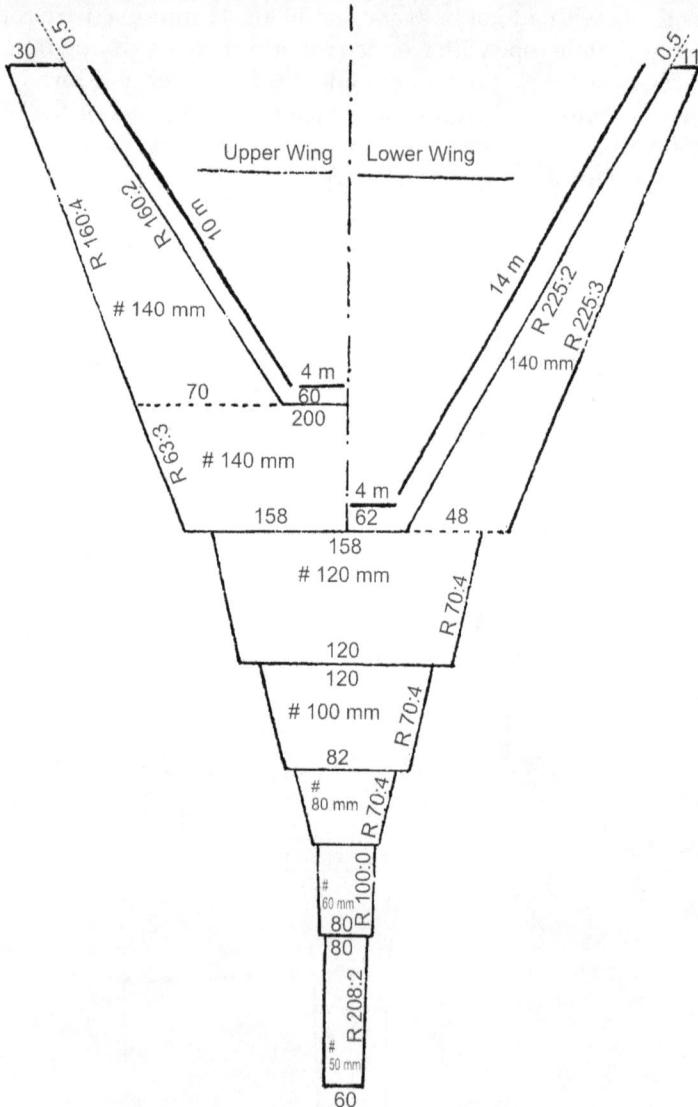

Fig. 77: Construction details of 24 m bottom trawl net.

35 meter trawl on the other hand was made up of 2 mm dia. monofilament twine weighing 42 kg. The mesh size was 25-70 mm bar. 8 mm dia wire rope wound with sisal twine of 36 m long served as head rope; while the foot rope was made up of 45 m long, 9 mm wire rope served with sisal twine. Bolch line was of 6 mm dia. monofilament rope. Aluminum alloy floats of 6 and 8 inch dia. with total buoyancy of 30 kg was used in the head rope of the net. The ground rope was made up of 11 mm dia. wire rope served with 51 mm dia. Manila rope with 4 inch dia. rubber discs. The hanging of the webbing was 50 percent for bosom and 12 percent each for upper and lower wing. The otter board area was 2 square meter weighing 200 kg. 40 meter long 9 mm dia. wire rope was used as sweep line. The construction details of 35 m bottom trawl is given in Fig.78.

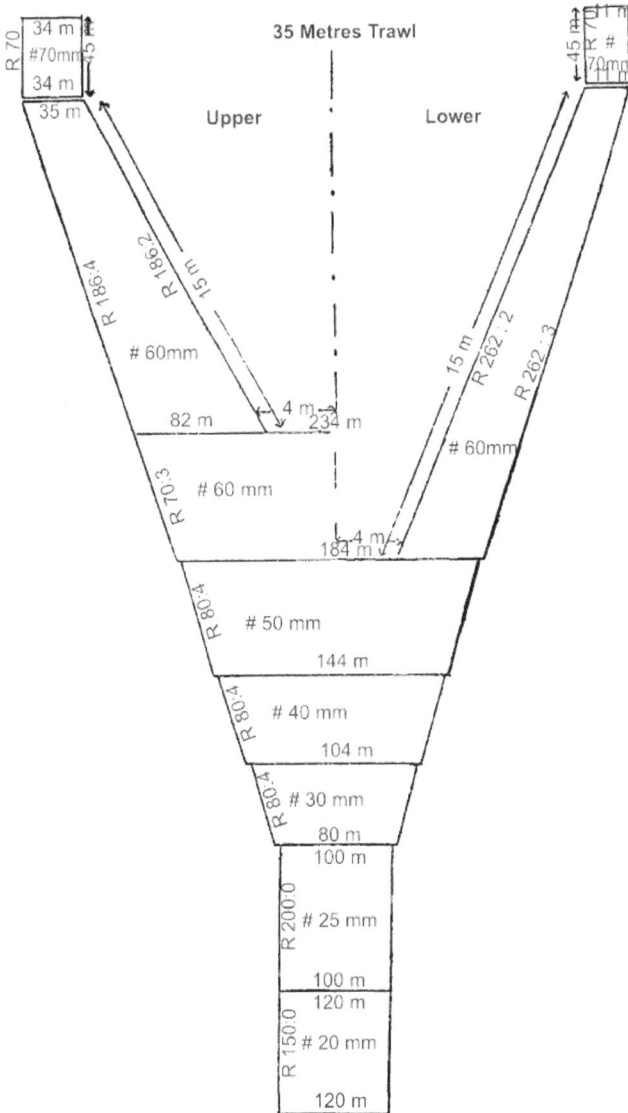

Fig. 78: Construction details of 35 m bottom trawl

The rigging details of these bottom trawls are given in Fig, 79

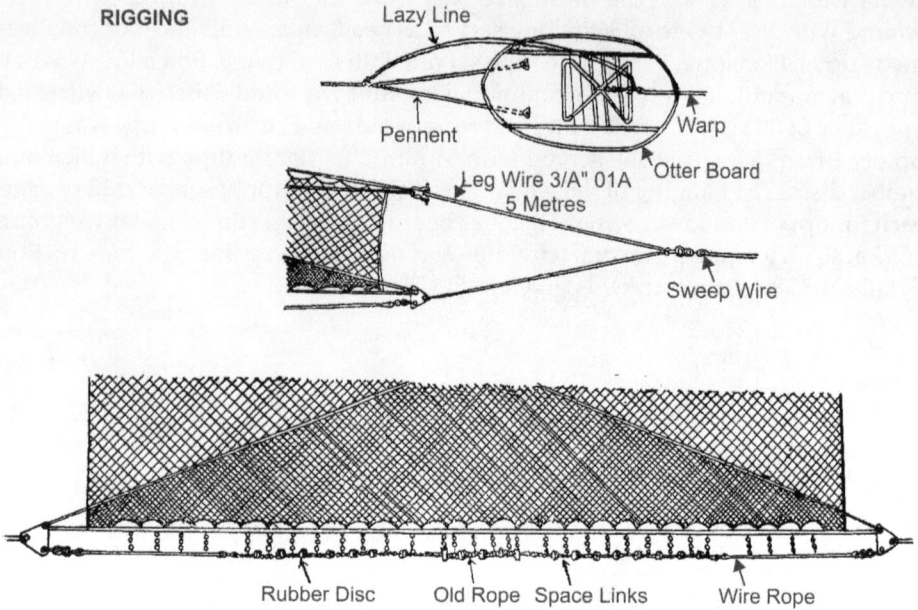

RIGGING

Lazy Line

Pennent

Warp

Leg Wire 3/A" 01A
5 Metres

Otter Board

Sweep Wire

Rubber Disc Old Rope Space Links   Wire Rope

Fig. 79: Rigging details of 15,24 and 35 meter bottom trawl.

Of the three trawls experimented, the 35 m monofilament trawl yielded the highest catch of 365 kg per hour of fishing effort followed by 205 kg per hour and 134 kg per hour in 24 m and 15 m trawl respectively from depths of 30 fathoms in Tamil Nadu coast in India.

51 M LONG WING SEMI-PELAGIC TRAWL

The technique of semi-pelagic trawling as an eco-friendly harvesting system is gaining momentum with its capability to exploit off-bottom resources that are generally not accessible to conventional bottom trawls. To enhance the effectiveness of pelagic and semi-pelagic trawls, incorporation of optimum mesh size in the fore part, lighter foot rope rigging, thinner gear material in the design and method approach along with employment of optimum towing speed, enhance the functional efficiency of semi-pelagic trawls. Absolute catching capacity of the trawl or the ratio of the amount of captured fish to the quantity available in the zone of operation for the trawling period rises with the increase of trawl size and towing speed (Freedman, 1971). Semi-pelagic trawls to be effective should develop an optimum vertical height and this can be achieved only with optimum wing length, towing speed and forked wing ends. Increase in the resistance of trawl under tow is proportional to the square of speed. (Jack Phillipes, 1959). Total drag and horizontal opening developed by the mid-water trawl or semi-pelagic trawl have definite correlation with speed of tow and the coefficient of correlation have been worked out. Towing speed being an important parameter an attempt was also made to arrive at an optimum speed of tow for semi-pelagic trawling in offshore waters. Manoharadoss *et al* (1999) have enumerated the effect of towing speed in inshore semi-pelagic trawling The versatility of "V" form doors provides scope for using these boards efficiently for semi-pelagic fishing.

The design details, rigging and functional characteristics of 51 m long wing semi-pelagic trawl based on fishing experiments conducted are described indetails. The experiments were carried out in the NW coast, off Saurashtra from a 71.5 m OAL 2285 HP during February 2000 with a 51 m semi-pelagic trawl having a pronounced lengthier wing intended to obtain an optimum vertical opening which is an absolute necessity for effective off-bottom functioning. The design details of trawl gear are shown in Fig. 80.

The net was uniformly fastened with 35 numbers of 200 mm diameter deep sea plastic floats (3.5 kg extra buoyancy each) along the head rope apart from 3 numbers of 300 mm diameter floats (4.5 kg extra buoyancy each) attached to each of the wing ends and the third to the center of the head line, to attain required extra buoyancy (136 kg) necessary to lift the net to an optimum level. 100 kg of 12 mm diameter GI link chain was attached to the foot rope in the method depicted in Fig. 80 (b). A front weight/depressor weighing 36 kg each (bunched 12 mm diameter GI link chain) was rigged to the distal ends of the foot rope, where it is connected to sweep lines. Increasing the vertical height of a variable depth trawl is effected by adding light double bridles and their length restricted to 50 m established to be the optimum. Field tests to assess the efficiency of the trawl gear in combination with perfect "V" form otter boards of dimension 2850 x 1800 mm weighing 1500 kg each were conducted to the area off Saurashtra coast. The position of the otter board with reference to the vessel was recorded with sonar and the speed of two was measured with Doppler speed log.

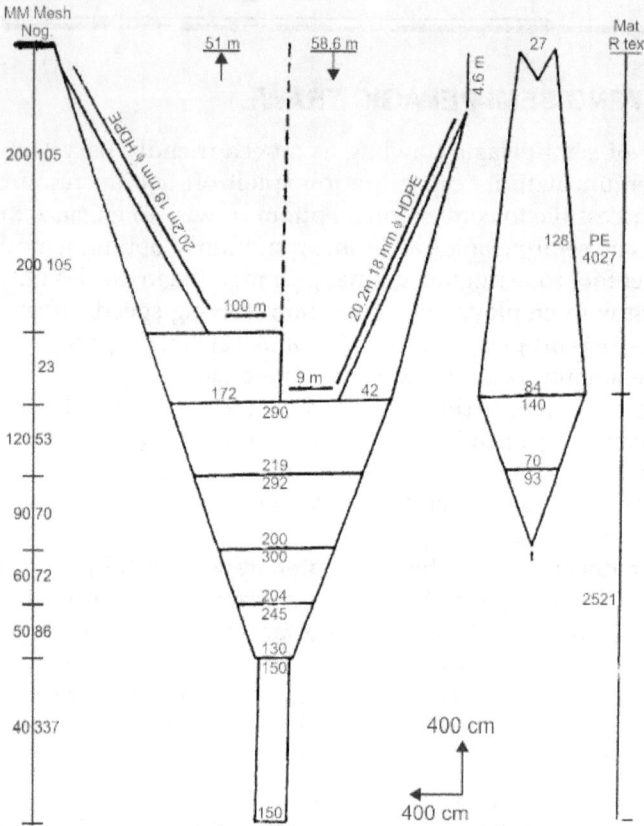

(a) Design details of 51 m semi pelagic trawl

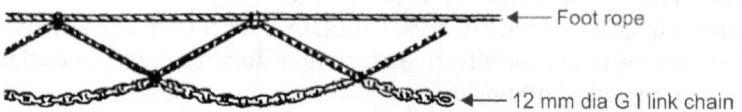

(b) Foot rope-ground rig
Fig. 80

A total yield of 3646 tonnes was realized from 24 hauls made and the average and maximum catch per unit of effort amounted to 123.6 and 151.0 kg respectively. 4 knot trawling speed was assessed to be the optimum for this gear system. More than 85% of the total catch contained semi-pelagic species of fish indicating the target specificity of the gear system developed. Commercially important species of fish consisting of *Tachysurus* sp, *Sphyraena* sp, *Trichiurus* sp, *Protonibea diacanthus*, *Pseudocaranx dentex, Rastrelliger kanagurta, Megalapsis cordyla, Polynemus indicus, Saurida tumbil, Chorenemus til, Chirocentrus* sp, *Scomberomoms* sp, squids and cuttle fish dominated the off-bottom resources caught.

It could therefore, be conclusively established that 51 m long wing semi-pelagic trawl designed and developed for operation from a larger class of vessel in combination with 2850 x 1800 mm "V" form otter doors and 50 m twin bridle system is suitable for off-bottom resources without detrimentally affecting the demersal ecosystem.

8

Shrimp Trawl

BELLY DEPTH FOR SHRIMP TRAWLS

In order to arrive the possibility of reducing the depth of belly of 17.07 m (56 feet) four seam trawl, when fishing for shrimps and also to evolve a possible mathematical equation to determine the maximum depth of belly in shrimp trawls, trials have been made at Cochin during the fishing season of 1967-68.

A 17.07 m (56 feet) four seam cotton trawl of non-overhang type with 140 meshes in its belly depth was chosen as the control net "A" to suit the capacity of the trawler. Two other experimental nets "B" and "C" were also simultaneously fabricated, which were identical to net "A" except for the number of meshes in the belly depth. The depth of belly in net "B" was 102 meshes, while that of net "C" was 80 meshes. In the net "B" the number of meshes reduced in the depth of belly was proportional to the number of meshes in 13.69 m four seam trawl, which is almost 2/5 or 40% of the stretched width of the belly along the bosom. The dimensions of the bellies in nets "B" and "C" are shown in the dotted lines of the figure 81.

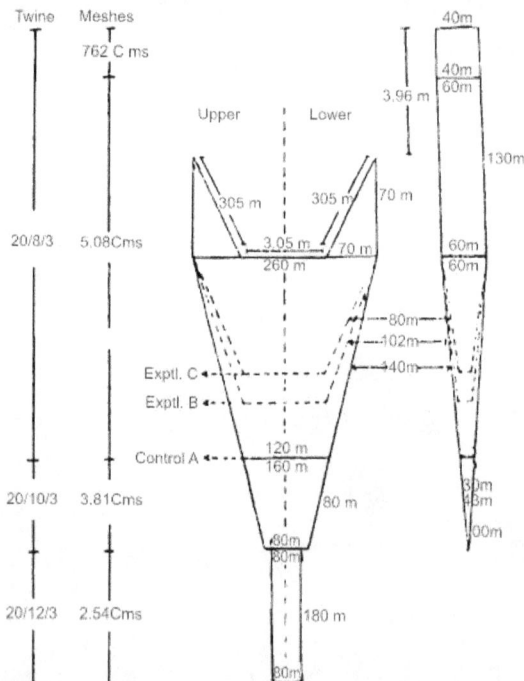

Fig. 81: Different stages of bellies and side wedges

All the three nets, "A", "B" and "C" were operated in rotation during each day of operation. To the best possible extent, all the fishing conditions, namely, fishing ground, fishing depth, length of warp released, towing direction, towing duration, towing speed, buoyancy of head rope, weight of sinkers on foot rope and the size of otter boards were kept constant.

In all 27 fishing trips were made and 27 hauls per net were taken on a comparative basis. Data regarding towing duration, percentage of horizontal opening, tension on warps (resistance) and catches collectively and individually were gathered and tabulated below.

Details of operations

Nets	No. of Hauls	Towing duration	% of horizontal opening	Tension on warp (kg)	Catches in kg. Prawn Fish Total
"A" 140 meshes	27	44min.	58.30	389.00	10.63 14.00 24.63
"B" 102 meshes	"	"	58.85	379.66	13.55 19.00 32.55
"C" 80 meshes	"	"	58.04	384.14	8.48 13,96 22.44

(Average of 27 operations for each net)

Earlier investigations had proved the possibilities of reducing the depth of belly for a 13.59 m (45 feet) four seam shrimp trawl. Based on the earlier findings and from the subsequent studies, it has been arrived at a mathematical equation, namely, $D = 2 L/5$, where "L" denotes the stretched width of the belly with 12 meshes for any marginal adjustments while either hand braiding or tailoring the webbing.

Further reduction of the belly depth resulted finally in the economy of the twine utilized, with least effect on the catching efficiency as well as the mechanical characteristics of the net. The reduction in the twine utilized was as significant as 47.9% and 27.6% in the case of 13.69 m and 17.07 m four seam cotton trawls respectively. This reduction also resulted in considerable reduction of the number of meshes in the belly region particularly and this accounted to about 46.16% and 27.785% respectively in the case of the above referred 13.69 m and 17.07 m trawls This adds to the savings of fabrication charges and there was definite reduction in the initial cost of fabrication of the gear on the whole.

9

Trawling Technology Transfer to the Users

Suitable trawl nets for medium sized trawlers

For diversified fishing up to 50 m depth suitable trawls to operate from medium sized boats have been designed and tried.

Three different design concepts of trawl nets, namely, bulged belly trawl, long wing trawl and four panel trawl, each having a head rope length of 32 m was employed in the trial. The design details of these three nets are given in figures 82, 83, and 84 and otter boards in Fig. 85.

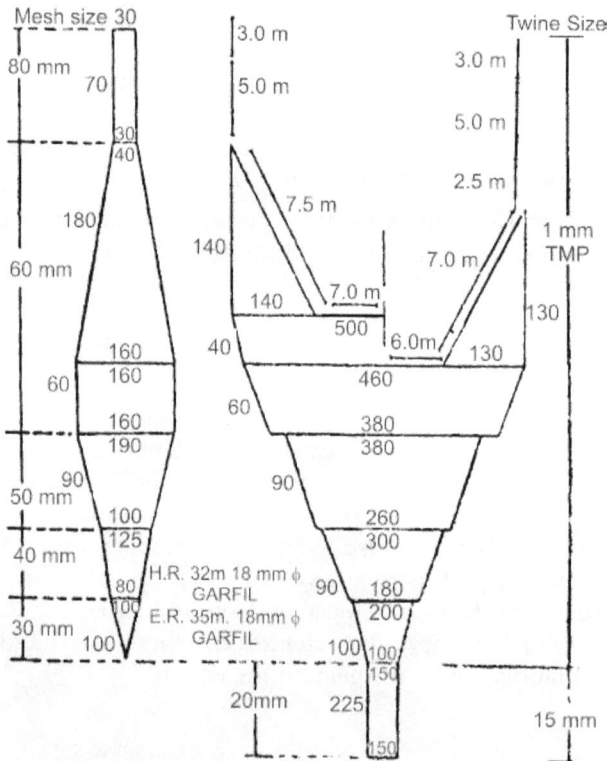

Fig. 82: Design details of 32 m bulged belly trawl

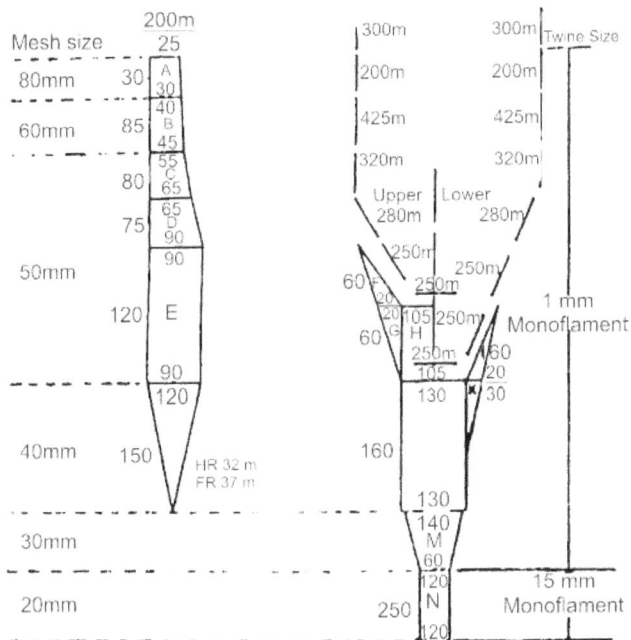

Fig. 83: Design details of 32 m long wing trawl

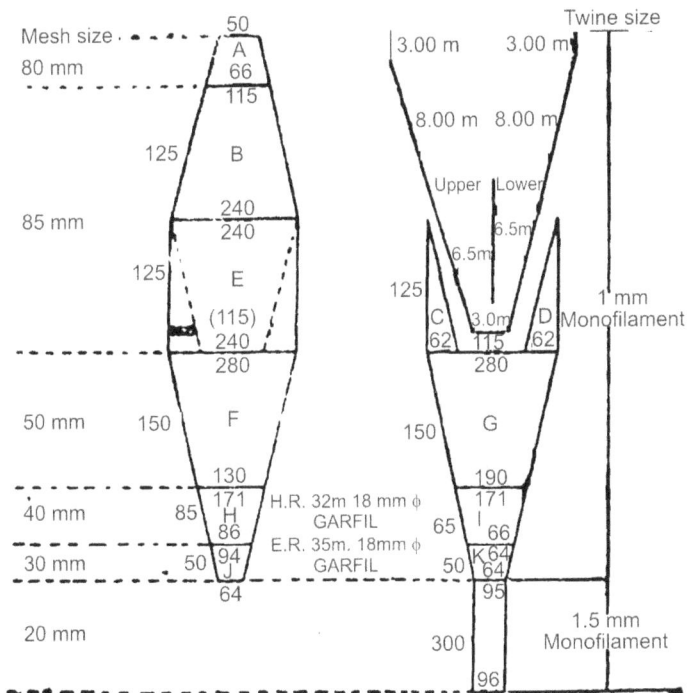

Fig 84: Design details of 32m four panel trawl

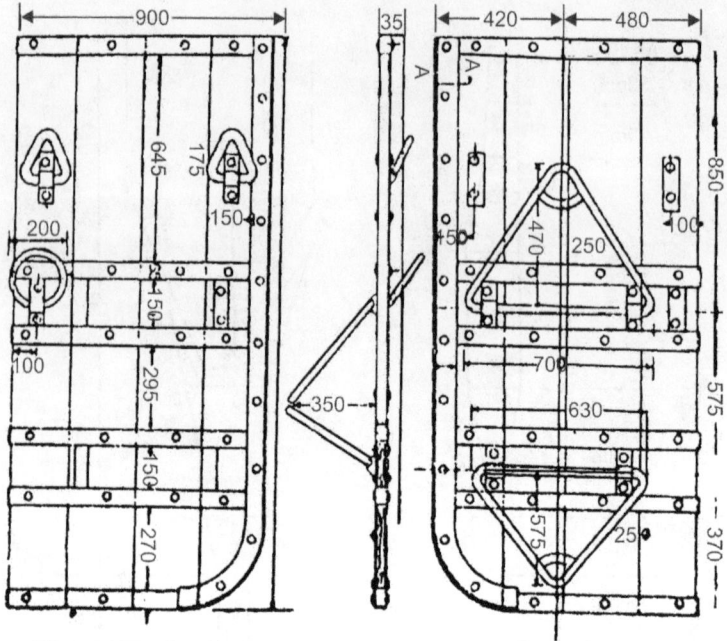

Fig. 85: Design details of otter boards rigged in 32 m trawls

The details of the nets, such as, requirement of twine, rope, number of meshes are given in the table below. Fishing was carried out from vessel of 15 m overall length. All the three nets were operated on each day giving equal chances for all the three nets. Parameters, such as, depth, length of rope, trawling speed, duration of each haul were kept constant for the three nets while fishing. Trials were carried out at depths 20-25 m, 25-30 m, 30-35 m, 35-40 m off Cochin from October to May of each year from 1972-1976

Details of three nets

Type of net	Total number of meshes	Qty.of rope and twine (kg)
32 m bulged belly	410000	Twine 30 kg
32 m long wing	215000	Twine-16, rope-12 kg
32 m four panel	390000	Twine-29 kg. rope- 12 kg

The data on the catch of prawn, column fishes and bottom fishes were recorded separately, depth wise for each net.

Catch details of three nets operated at different depths.

Depth (m)	Hauls (nos)	Bulged belly trawl				Long wing trawl				Four panel trawl			
		A	B	C	D	A	B	C	D	A	B	C	D
20-25	24	44	12	200	256	111	8	121	240	33	34	205	272
25-30	54	39	5	963	1007	55	1	1140	1196	24	4	839	867
30-35	36	30	486	396	912	48	152	483	683	20	424	220	664
30-35	24	nil	183	614	797	nil	29	546	575	nil	95	366	461

35-40	45	nil	39	1084	1123	nil	22	725	747	nil	25	608	633
Total(kg)	113	725	3257	4095	214	212	3015	3441	77	582	2238	2897
Catch/hour				66				56					47
Catch (%)	28	48	40	40	52	14	35	33	20	38	25	27

A = Prawn (kg); B = Column fish (kg); C = Bottom fish (kg); D = Total (kg)

The bulged belly trawl was found to be more efficient *in* comparison with the other two, so far as catching rate and other performance are concerned. Analysis of prawn catch at 20-25 m and 25-30 m depth have shown that the long winged net to be more efficient for prawn compared to other two nets. The efficiency of long wing trawl can be attributed to its specific design, such as, extra long wings sweeping more areas of the sea bed while trawling. 47% of the total column fishes and 40% of the total bottom fishes were landed by bulged belly trawl. The long wing trawl caught more of bottom fishes (35%) and in this respect was superior to the four panel one which accounted for only 25% of the total bottom fishes, while in the case of column fishes four panel was better than long wing trawl as it caught 38% of the column fishes. The four panel net also caught more column fishes but found to be next to bulged belly. It is concluded that the long wing trawl, which is 50% cheaper than other two trawls can be employed up to a depth of 20 m for prawns. The bulged belly can be recommended for deeper waters.

Design aspects of four seam trawls

To find out the changes that have taken place in the design aspects of trawls since their introduction, the studies relating to certain aspects of the design of four seam trawls operated from Tuticorin in the East coast and Mangalore in the West coast have been undertaken. The size of vessels used for trawling ranged from 30 feet 30 HP wooden boats to 93 feet 600 HP steel trawlers, and the size of nets ranged from 15 m to 55 m. These gears fall under 30 different designs of 4 seam trawls and 14 two seam trawls.

It is an established fact in designing trawls, that it has to be designed so as to match the resistance of the net with the pull of the boat for the effective utilization of available engine power. Even though there may be diversity in the design of trawls, there exists certain inter-relationships between the different parts of the trawl and its size. Miyamoto (1959) has shown that the size of the net can be expressed as a function of the horse power of the engine. From the survey, the relation between the length of head rope (HR) and horse power (P) of the engine can be expressed as follows.

$$HR = 0.0615\,P + 16.4244\ (m) - 1\ (Fig.\ 86)$$

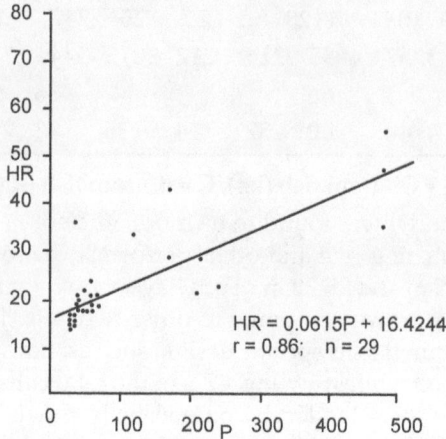

Fig. 86: HR = 0.0615P + 16.4244 ; r = 0.86; n = 29

The relation between the maximum belly width (L) and the head rope length can be derived as;

$$L = 0.4454\,HR + 6.433\,(m) — —2\,(Fig.87)$$

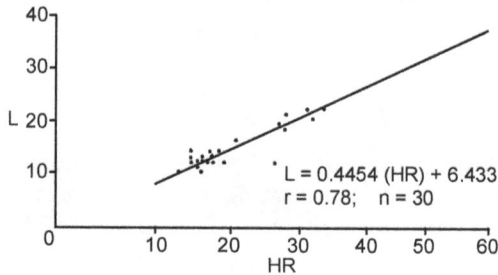

Fig. 87: L = 0.4454 (HR) + 6.433 ; r = 0.78; n = 30

The other relationships worked out in relation to the maximum width of belly are, depth of belly (D), length of Bosom (Bm), minimum width of belly (B), width of side wedge (W) and height of jib (J). These relationships can be expressed in the following equations.

$$D = 0.7426\,L + 0.32\,m -—3\,(Fig.\,88)$$

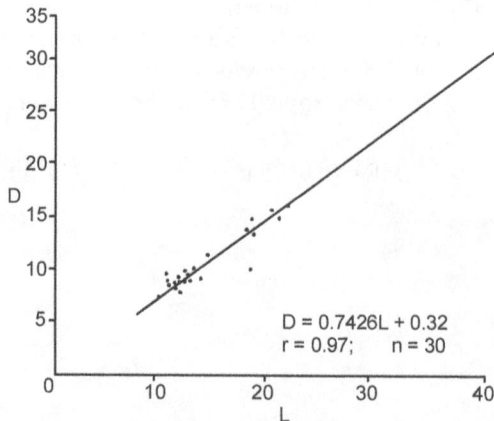

Fig. 88: D = 0.7426L + 0.32; r - 0.97; n = 30

$$Bm = 0.4706L\text{-}0.472 \text{ m-4 (Fig. 89)}$$

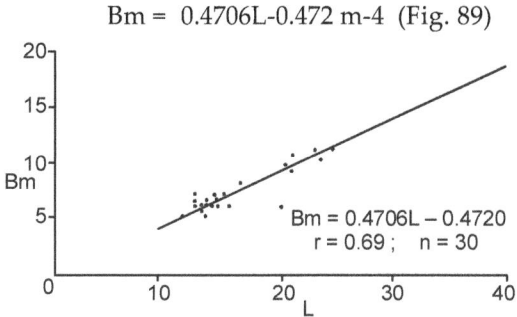

Bm = 0.4706L – 0.4720
r = 0.69 ;   n = 30

Fig. 89: Bm - 0.4706L - 0.4720; r - 0.69; n = 30

$$B = 0.1616 \text{ L} + 0.9820 \text{ m} - 5 \text{ (Fig. 90)}$$

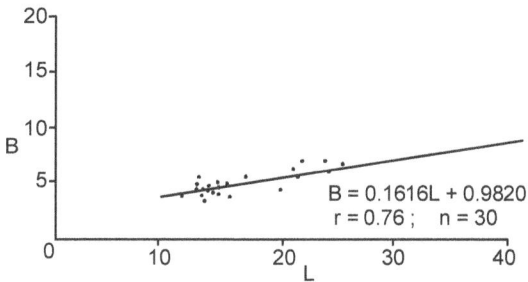

B = 0.1616L + 0.9820
r = 0.76 ;   n = 30

Fig. 90: B = 0.1616L + 0.9820; r - 0.76; n = 30

$$W = 0.25 \text{ L} + 0.3187 \text{m} - 6 \text{ (Fig. 91)}$$

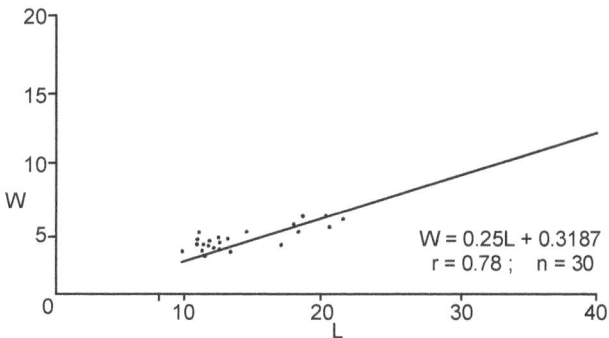

W = 0.25L + 0.3187
r = 0.78 ;   n = 30

Fig. 91: W = 0.25L + 0.3187; r = 0.78; n = 30

$$J = 0.3307L = 0.131 \text{ m} - 7 \text{ (Fig. 92)}$$

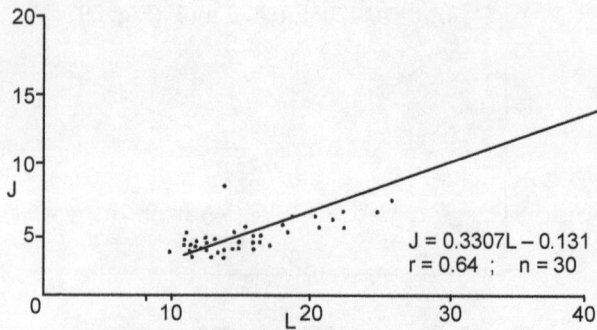

Fig. 92: J = 0.3307L-0.131;r = 0.64;n = 30

The relation between the extra buoyancy (F) of floats to the head rope has been worked out as;

$$F = 1.2514 \, HR - 9.8961 \, (kg) - \sim 8 \, (Fig. \, 93)$$

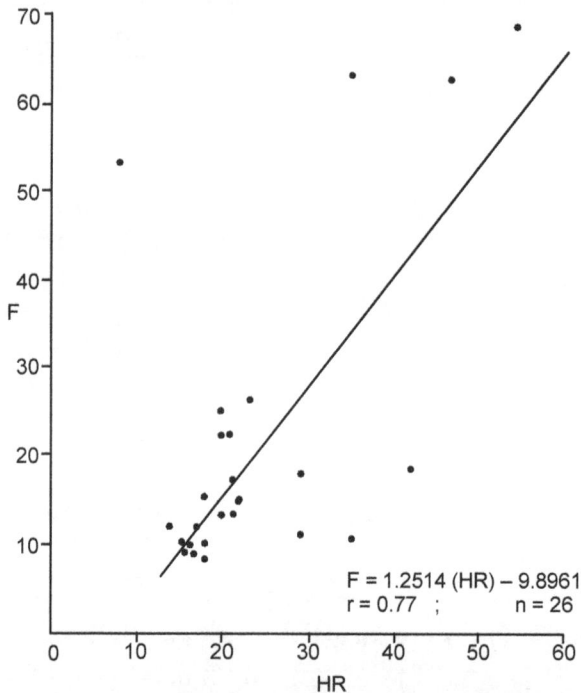

Fig. 93: F - 1.2514(HR); r = 0.77; n = 26

The weight of sinkers (Wt) in relation to the foot rope (FR) can be expressed as;

$$Wt = 2.0735 FR - 21.4321 \, (kg) \, 9 \, (Fig. \, 94)$$

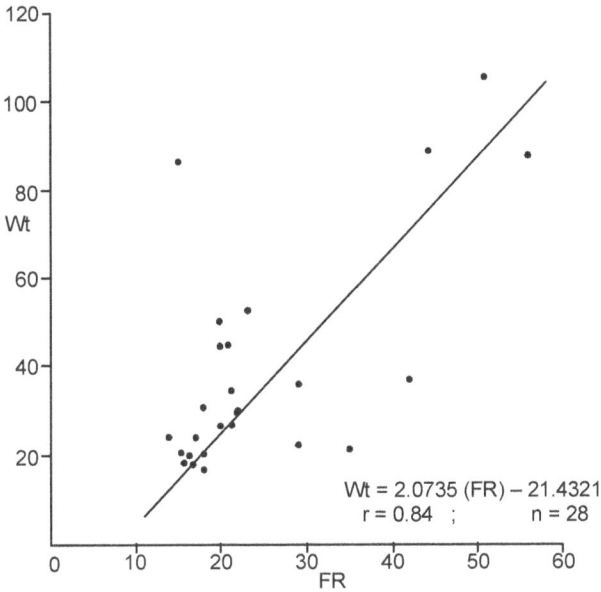

Fig. 94: Wt = 2.0735(FR) - 21.4321; r = 0.84; n = 28

The analysis of the data has indicated that in addition to the above, certain other relationships between the various parts of the net in relation to Head rope length, can also be formulated. These relationships are expressed in the following equations.

$$LW = 0.375 \text{ HR-1.81m} - 10 \text{ (Fig. 95)}$$

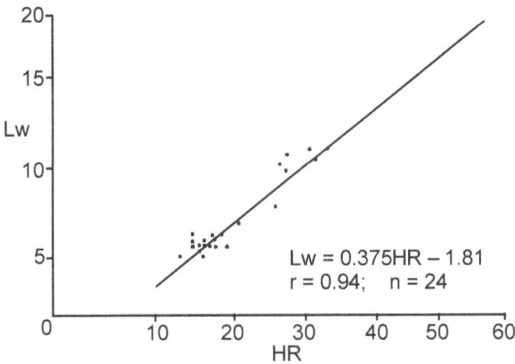

Fig. 95: LW = 0.375HR - 1.81; r = 0.94; n = 24

$$C = 1.0163 \text{ HR} + 17.6486\text{m} - 11 \text{ (Fig. 96)}$$

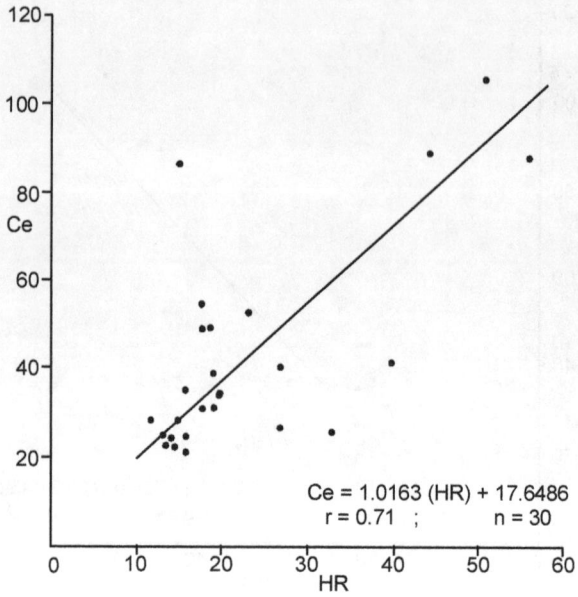

Fig. 96: Ce = 1.0163(HR) + 17.6486; r = 0.71; n = 30

$$OAL = 0.84\ HR + 9.3\ m - 12\ (Fig.\ 97)$$

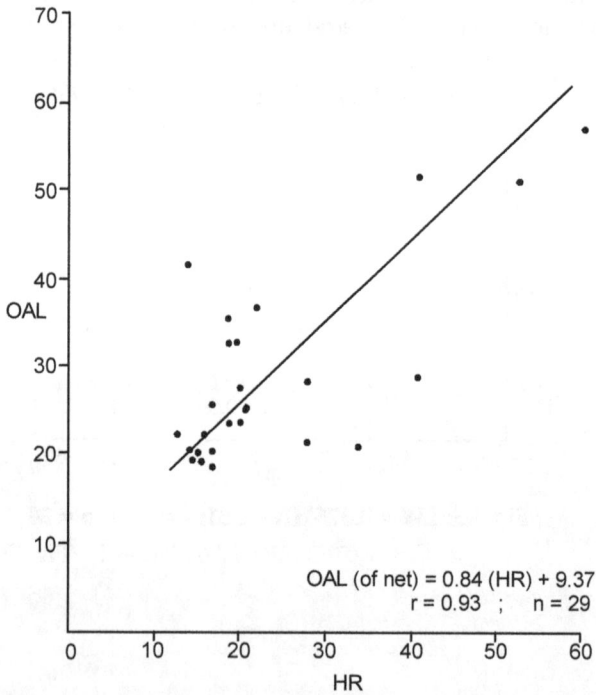

Fig. 97: OAL (of net) - 0.84 (HR) + 9.37; r = 0.93; n = 29

$$LCE = 0.0896\ HR + 3.5644\ m - 13\ (Fig.\ 98)$$

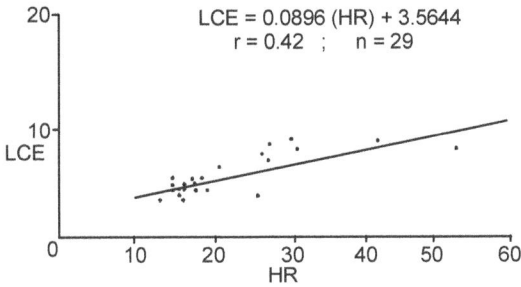

Fig. 98: LCE = 0.0896(HR) + 3.5644; r = 0.42; n = 29
Area of otter board = 0.0247 HR + 0.4950 sq. m-14 (Fig. 99)

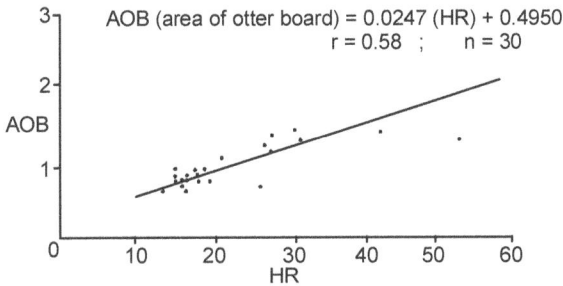

Fig. 99: AOB (area of otter board) - 0.0247(HR) + 0.4950; r = 0.58; n = 30;

Where LW = Length of wing; C = Circumference of the mouth of the net; OAL = Overall length of the net from tip of wing to cod end and LCE = Length of cod end

The critical analysis of the present day relationships will indicate certain aspects. The initial attempts of analysis of the design details of four seam trawls operated off Cochin area, with a view to establish similar relationships made by Miyamoto (1958). Nair and George (1964) has also conducted similar survey and arrived at certain relationships. A comparison of the relationships established now and in 1964 shows certain interesting facts. The respective equations arrived are detailed below.

Sl.No.	In 1964	In 1970
1.	HR = 0.5 P+28.1(feet)	HR = 0.0615 P+ 16.4244 m
2.	L=0.7HR + 6(feet)	L = 0.445 HR + 6.433 m
3.	D = 0.7 L +2.3 (feet)	D = 0.7426 L + 0.32 m
4.	Bm = 0.5 L - 2.87 (feet)	Bm = 0.47o6 L - 0.472 m
5.	B = 0.24 L + 0.92 (feet)	B = 0.1616 L + 0.9820 m
6.	W = 0.4L-4(feet)	W = 0.25 L + 0.3187 m
7.	J - 0.22 L +2.46 (feet)	J = 0.3307 L-0.131 m
8.	F = 0.75HR-15(lbs)	F= 1.2514 HR-9.8961 (kg)
9.	Wt = 0.8 FR- 9.65 (Ibs)	Wt-2.0735 FR-214321 (kg)

An average increase of 4% in the size of trawl (head rope length) has been observed in the present day design. This may be attributed to the decrease in the diameter of twine used for the fabrication of the nets. It is also seen that there is an average 4% decrease in the minimum width of the belly and a 2% increase in the depth of the belly.

It is very likely that this is due to the combined effect of different tapering ratios adopted for stream streamlining the tapering edge of the belly. It will be evident from the above table that there has been a 6.6% increase in the width of the side panel and 26.8% increase in the buoyancy of the float. This trends show a more vertical opening in the designs. The weight of sinkers was found to be increased on an average 7.4%, which probably would have become necessary to compensate for increased buoyancy of floats used.

Trawling technology transfer and innovation decision of the end users

Innovation decision is a process through which an individual or other decision making unit passes from first knowledge of an innovation ,to a decision to adopt or reject, to implement the new idea and to confirmation of this decision.

Research institutes have transferred several of its innovation to various categories of users over the years. The study of innovation decision process include (a) to study the socio-personal and psychological characteristics of fishing trawler operators; (b) to examine the efficiency of innovation decision process among them; (c) to study the relationship of their socio-personal characteristics with innovation decision efficiencies and (d) to document the constraints.

Out of eleven identified technologies, V- form steel otter boards, large mesh trawl, turtle excluder device (TED), square mesh cod end trawls, , appropriate engine horse power of the boat, trawl designs used (Size/type) relates trawling technology. The innovation decision efficiency index (IDEI) of a respondent was calculated for each of the above technology as;

IDEI = Total score of an individual for passing through the five stages (namely, knowledge, persuation, decision, implementation and confirmation)

$$----------------------------------X100$$

Maximum score

The socio-personal and psychological characteristics of the respondents indicate that nearly equal proportion of the fishing trawler operators belonged to old and middle aged groups (50 and 49% respectively). More than one third of them (35%) was educated up to primary school level. Nearly two-third of the fishermen (63%) belonged to nuclear family system with a little more than one-third (36%) belonging to joint family system. The mean family size was five. A vast majority of the respondents (91%) had fishing along as their primary occupation, only 9% of them had other occupations in addition to fishing.

The mean investment on a fishing unit was 15.19 lakh rupees, which comprised of one fishing craft, 8 to 10 fishing nets and one engine (110-140 HP) and for electronic instruments used in the fishing vessels, such as, Geographical Positioning System (GPS), echo-sounder and wireless transmitter. The average experience of the fishermen in fishing was 17 to 31 years. In a year the fishermen were spending only 188 days in fishing due to the mandatory ban on fishing for 45 days during the breeding season enforced by the state department, rough seas, climate factors, availability of resources, lay-off during festive seasons, frequent repair and maintenance works and the ever increasing operational expenditure limiting their days of fishing.

The mean and annual income (Rs in '000) of the respondents was 97.86 plus minus 38.50. The average annual expenditure on repair and maintenance of fishing crafts and gears was (Rs in '000) 158.86 plus minus 57.62.

The majority of the respondents (64%) was under medium category followed by high (27%) and low (8%) in their level of innovativeness. 70% of them were possessing medium level of economic motivation followed by high (20%) and low (9%) levels. It was also observed that almost equal percentages of respondents belonged to low (40%) and medium (37%) categories in their social participation. High social participation was observed among one-fifth of them only. Nearly 60% of the mechanized boat operators had low level of extension participation. Majority (90%) of them had low level of exposure to training program.

Majority of the respondents rarely utilized the formal source of information, such as, researchers (80%) and NGO officials (77%) though more than half of them (55%) had occasionally used extension personnel for receiving technological information

The overall innovation decision efficiency index score was 79. The index score on passing through the five stages of innovation decision process, namely, knowledge, persuasion, decision, implementation and confirmation were 96, 82, 72, 72, and 72 percentages respectively.

In the case of Turtle Excluder Device (TED) the mean innovation decision efficiency index score was 23%. The results indicated that even though more than 75% of the fishermen knew about the existence of the innovation, only about one-third of them passed through the persuation stage. This is due to the fact that they have not agreed with the attributes of this innovation and formed an unfavorable attitude towards this innovation. The index scores on passing through the decision implementation and confirmation stages were zero each. This might be due to the fear that the use of the innovation would result in catch losses. Since the decision stage is very crucial, the fishermen could be convinced to take a decision to adopt this innovation through the use of various extension methods. CIFT studies have established that the overall catch losses during trawling operations from commercial fishing vessels, due to installation of TED had been 0.5% for shrimp and 2.75% for non-shrimp catch component.

In the case of the square mesh cod end trawls, the mean innovation decision efficiency index score was 26%. The results indicated that a vast majority of the fishermen knew about the existence of the innovation, how it functions and its consequences. Only about one-third of them passed through the persuation stage and the remaining skipped it. The index scores on passing through the decision, implementation and confirmation stages were zero each, which indicated that none of them were passing through these three stages. This might be due to the fear that the use of the innovation would result in catch reduction. Though CIFT studies showed that cod end mesh size of 30 mm in demersal trawls provided better chance of escapement of most of the juveniles, while retaining bulk of commercially important species. Since the decision stage is very crucial, the fishermen could be convinced to take a decision to adopt this innovation through the use of various extension methods, in the interest of conservation of juveniles to ensure sustainable fisheries.

The innovation decision process was efficient pertaining to the technologies which were directly related to increasing production, labor efficiency, fuel efficiency, reducing the operational expenditure and increasing the income. In the case of technologies pertaining to the conservation of resources in the interest of sustainability and environmental impact, the innovation decision efficiency scores were relatively low. However it could be concluded that overall innovation decision efficiency pertaining to the selected eco-friendly and responsible technologies was fairly good among the respondents of the fishermen of Ernakulam district of Kerala state engaged in marine fishing.

It could be understood that the fishermen group required more information regarding the practicability, feasibility and the cost-benefit ratio of technologies in evaluating the technologies in their innovation decision behavior. The mass media and extension agencies have vital roles to play to bridge the gap to pass through different stages. At persuation stage, they could be motivated to form a favorable attitude towards the innovation, as they are more psychologically involved with the innovation. Passing through the decision, implementation and confirmation stages is the fishermen's choice, as extension agents have minimum roles to play at these three stages. However, the extension agents can provide opportunities to the fishermen to witness the advantages of innovation to ensure the availability of technological inputs and resources to put the innovation into practice and to see that the fishermen are not exposed to conflicting messages about the innovation.

Lack of training and access to research and extension system, lack of information on technologies, increase in cost of inputs, lack of financial resources, increase in operational expenditure, lack of infrastructural facilities, non-availability of inputs or resources and diminishing resources were perceived as constraints in the innovation decision process pertaining to most of the identified technologies.

<table>
<tr><td>10</td><td># Trawl Selectivity on the Catch</td></tr>
</table>

Trawl selectivity on the species

In order to ascertain the trawl cod end selectivity on sciaenids *Johnius* spp. in Gujrat coasts, India, where trawl fishery contributed 61.36% of total catch, fishing trials with a 34 m high opening bottom trawl was carried out from 15.5 m LOA, 124 HP stern trawler. Due to the use of small sized meshes ranging from 15-30 mm, in the cod end, the trawlers along Gujrat coast generate disproportionately large quantities by - catch which also includes juveniles of commercially important species.

The knowledge of selectivity of commercially important gears is vital for effective monitoring, management and sustainable exploitation of fishery resources. Use of selective gears helps to minimize the capture of juveniles by regulating the length at first capture, thereby increase the yield per recruit of the targeted species and also help in reduction of discards in the fishery. Among different technical measures attempted to improve the selectivity of trawl gears, changes made to the shape and size of cod end meshes was found to be the most adoptable due to its inherent simplicity and easiness to modify.

Trawl selectivity experiments are often carried out by covered cod end technique and the selectivity estimates derived from the escapement and retention data of the species from multiple tows, lead to variance in the selection parameters need to be taken into account to make the selectivity estimates more reliable. Over-dispersion in the data can be quantified using the replication estimate of dispersion.

Selectivity experiments with respect to *Johnius dussumieri* carried out using 40 mm diamond mesh cod end in a trawl net operated off Veraval, Gujrat. A 34 m high opening bottom trawl (HOBT) rigged with 40 mm diamond mesh cod end was used for the selectivity experiments.

The wings to the square of the net were fabricated using 200 mm meshes and the length of the cod end was 8 m with a circumference of 170 meshes. "V"-form otter boards with dimensions of 790 x 1300 mm weighing 85 kg each were used with the experimental trawl. Cod end with 40 mm diamond mesh was constructed with high density polyethylene (HDPE) netting of 1.5 mm twine size. The cover for the cod end was fabricated using 20 mm (Rtex 630) polyamide netting and proportionately 1.5 times longer and larger than the cod end to minimize the masking of cod end by cover. All the trawling operations were carried out during day time and identical shooting and hauling procedures were adopted during the entire fishing operation. The duration of single tow varied from 1.5 to 2 hours and the speed of trawling varied from 1.03 to 1.13 meter per second. At the end of each tow, the catches from the cover and the cod end were separately sorted, individual species were weighed and the total length (TL) measured to the nearest 0.5 cm.

"Stacked haul method" (Millar *et at,* 2004), which accounts for the between-haul variation by implicitly keeping the replications of length classes from all hauls separately and allows the estimation of replication estimates of dispersion (REP) to be estimated was used in the estimation of variance. Data from all the hauls were stacked, into a single data set, which was then handled as a single (artificial) haul (Macbeth, et al, 2007). Scaling of data was carried out wherever necessary. Logistic selection curve was fitted to the stacked data for each species cifit (Millar, 2006). The standard error (SE) of all the estimates was REP corrected for the between haul variations. Selectivity parameters were estimated using the coefficient "a" and "b" derived by maximum likelihood method (Wileman *et al,* 1996). The 50% retention length of a species was calculated as L 50 = a/b, selection range (SR) = L 75 - L 25, selection factor (SF) = L50/mesh size and selection ratio (SRA) = SR/L50. Model fits were assessed by comparing REP-corrected deviances and associated degrees of freedom (d') against chi distribution and the appropriate model was selected (Macbeth *et al,* 2007). For calculation of the SF and SRA, the nominal mesh size of 40 mm was used.

Cod end escapement data of *J, dussumieri* collected front 6 hauls using 40 mm diamond mesh cod end reveals that the size selection curve for the species in 40 mm diamond mesh cod end along with the population retained by the cod end cover. The total number of individuals encountered in the net during the operation was 696, the major portion of the population encountered was to the left of the line indicating Lm50. The length of 50% retention (L50) + SE of the species was noticed as 7.55 (plus minus 0.12) cm .The selection range was worked out as 3.41 (plus minus 0.18) cm. The selection factor and selection ratio were 0.19 and 0.45 respectively. A total of 310 individuals were retained in the cod end, while 386 individuals escaped to the cover. Considering the length of first maturity to be 122 mm, the optimum mesh size required in the cod end (diamond mesh) for conservation of the species was estimated as 64.2 mm,

The currently used cod end mesh size in trawl fishery (ranging from 15-30 mm) is grossly insufficient for the conservation of the species. The age of capture of the species being 0.676 years and significant quantities of juveniles of this species have also been encountered in the by – catch and from the estimates of selection factor, it can be concluded that the cod end mesh size of 65 mm can be used for conservation of *J. dussumieri* along the Gujrat coast, based on the assumption on that 50% of the individuals would be retained as breeding biomass. However, it is in general agreement with other findings, that square mesh cod end are more selective than the diamond mesh cod ends of the same size.

Trawl selectivity in respect of silver pomfret

Tropical trawl fisheries produce large amount of by catch containing various aquatic organisms including juveniles of commercial fishes. Responsible fishing requires that, fishing gear should preferentially catch the adult fish at a particular age which would maximize yield while permitting the juveniles and sub-adults to escape and also minimize the catch of non-targeted and protected organisms. In this context, studies on cod end selectivity assume importance as it would facilitate choice of mesh size to selectively harvest the target fish of particular size which would have spawned at least once to ensure long term sustainability of the fishery resources.

Selectivity studies using covered cod end method were carried out from a 17.5 m LOA, 278 HP vessel off Cochin during 2007 using 18 m semi-pelagic trawl fitted with 100 mm diamond mesh cod end made of knotted polyethylene netting with 1.5 mm diameter twine. The cod end was provided with a cover made of 30 mm mesh size polyamide netting with 210 d x 2 x 2 twine which is approximately 1.5 times the size of the cod end. During selectivity experiments hauls of one hour duration each were carried out at a depth range of 10-25 m at a trawling speed of 2.3 to 3 km. The *Pampus argenteus* retained in the cod end and cover were separately collected and the fork lengths (FL) were measured to the nearest mm. The retention probability of pooled data from multiple hauls was modeled by means of logistic selectivity curve.

Selectivity parameters of 100 mm diamond mesh cod end in respect of *Pampus argenteus* based on the results of selectivity experiments are shown below;

Selectivity parameters	Value
L50(FLmm)	147.8
Selection range (mm)	27.3
Selection factor	1.45
Length at first maturity female (FL, mm based on Ghosh *el al*, 2009)	217
Recommended cod end mesh size (mm)	150

Total length (TL) and standard length (SL) were converted to fork length (FL) using a regression equation given by Froese and Pauly (2011).

Conversion equation:

$$TL = 0 + 1.265 \times FL; TL = 0 + 1.407 = SL; SL - 0 + 0.899 = FL$$

The mean selection length is generally proportional to the mesh size of the cod end over a certain range (Gulland, 1969; Fryer and Shepherd, 1996). The optimum mesh size was estimated as 150 mm based on the highest reported value of length at first maturity for silver pomfret in Indian waters and the selection factor determined through trawl selectivity experiments.

Information on trawl selectivity is essential in biological investigations, fish stock assessment, fisheries management and fishing gear design and development. Based on the results of selectivity experiments, the mesh size that can be recommended to harvest silver pornfret is 15.0 cm for conventional diamond mesh cod ends in order to protect juveniles and sub-adults.

Square mesh cod end selectivity for *Saurida tumbil* and *Nibea maculate*

Saunda tumbil and *Nibea maculate* are two commercially exploited species along the east coast of India. Large quantities of juveniles of both the species are landed at

Visakhpapatnam, in trawl nets with conventional cod ends having 10 to 20 mm diamond mesh. The use of diamond mesh leads to narrowing at the middle of the cod end causing the mesh lumen to almost close during trawling and hence smaller fishes are retained in the cod end.

Cod end selectivity plays a very important role in minimizing the capture of juveniles by regulating the size at first capture, increasing the yield per recruit of targeted species, reducing the discards and hence the impact of fishing on ecosystems. Knowledge on selectivity of cod end mesh size for all the commercially important species in a given fishing area, is of great significance in determining the judicious exploitation of fish stocks. The size selectivity of the trawl cod end is primarily affected by mesh size and by other factors such as mesh shape, twine characteristics and net construction. Mesh selectivity studies on commercially important species are essential to identify the fishable segment of the stock.

The shape of the cod end effects the selectivity of cod ends and the superiority of square mesh has been proven by many workers (Robertson et al.,1986).

The size selectivity of 40 mm square mesh cod end for *S. tumbil* and *N. maculate* was studied with a 30 m demersal trawl following covered cod end method.

The L25, L50 and L75 values for *S. tumbil* with 40 mm square mesh cod end was 16.2, 19.3 and 22.5 cm respectively.Selection range, selection factor and selection ratio for *S. tumbil* were 6.2, 4.8 and 1.5 respectively. The L25, L50 and L75 values for *N. maculate* with 40 mm square mesh cod end was 9.3, 10.9 and 12.5 cm respectively. Selection range, selection factor and selection ratio for *N. maculate* were 3.2, 2.7, 0.8 respectively. Characteristic smooth sigmoid curves typical for towed gears were obtained for *S. tumbil* and *N. maculate,* As size increases, the percentage of fish retained also increases until escapement is zero and all fishes are retained. When compared to square mesh, diamond-shaped mesh elongate under tension. During hauling, as the cod end fills with fish, the end meshes are obstructed, water flow is diverted and the cod end becomes bulbous. Fish escape mainly through the open meshes at the front of the bulb, while forward of this, most meshes are stretched and closed. In contrast, square-shaped meshes remain open during towing and do not acquire a bulbous shape. Square meshes have generally been more selective than diamond shaped mesh sizes for round fishes, like, haddock and whiting and hake. In the case of flat fishes where selection is related to width rather girth of fish, square mesh cod end is seen to be less effective in releasing young ones.

The length of first maturity (LFM) *of S. tumbil* in Indian waters is reported to be 29.5 cm. (Rao, 1984). The L50 derived from the study for *S. tumbil* in 40 mm square mesh cod end, was 19.3 cm, which is lower than the length at first maturity values reported. The optimum mesh size derived from the length at first maturity of *S. tumbil* (19.3 cm) and the selection factor of 40 mm square mesh determined through trawl selectivity experiment was 62 mm.

The LFM of *N. maculate* is reported to be 18.5 cm. The L50 of *N. maculate* derived for 40 mm square mesh cod end was 10.9 cm which is lower than the reported LFM of this species. The optimum mesh size for *N. maculate* determined from LFM and selection factor for 40 mm square mesh was 68 mm. Based on the results of selectivity experiments, the mesh sizes for square mesh cod ends, that can be recommended to harvest *S. tumbil* and *N. maculate* are 62 mm and 68 mm respectively in order to protect the juveniles and sub-adults.

Trawl cod end selectivity of Torpedo Scad, *Megalaspis cordyla*

The torpedo scad, *Megalaspis cordyla* (Linnaeus, 1758) is a moderately large marine fish belonging to family Carangidae. The largest recorded individual was 80 cm long (TL). It is a high level carnivore with a trophic level of 4.40 and feeds on a variety of fish, cephalopods and crustaceans. The torpedo scad is distributed throughout the tropical and sub-tropical waters of the Indian and West Pacific Oceans.

Torpedo scad is an important pelagic fish resource in India. Average landings during 2009-10 period was about 32000 t which formed over 18% of the carangid landings and 1% of the total marine landings in India. The size range in commercial landings is 20.0-67.0 cm and the species grows to 21.6, 35.8, 43.8, 48.5 and 51.5 cm in 0.5, 1, 1.5, 2, and 2.5 years respectively. The species is caught mainly in trawls, drift gill nets, purse seines and hook and lines in Indian waters.

Selectivity studies using covered cod end method were carried out onboard research vessels LOA 15.24 m, 223 HP and LOA 17.5 m; 278 HP, off Cochin during 2007–2008, using 18 m semi-pelagic trawl, fitted with 65 mm diamond mesh cod end. The cod end was provided with a cover made of 30 mm mesh size polyamide netting, which is approximately 1.5 times the size of the cod end as suggested by Stewart and Robertson (1985). During the selectivity experiments, 45 hauls of I hour duration each were carried out in the depth range of 10-25 m, at a trawling speed of 2.3-3 knots. Samples were drawn from the cod end and cover and the length frequency data were recorded for the selected species, torpedo scad (*Megalaspis cordyla*). The logistic model commonly used to describe trawl selection ogive (Sparre et al, 1989) was adopted for the study.

According to the selectivity studies, the L25, L50, and L75 values for *Megalaspis cordyla* were 231, 294 and 357 mm respectively. Selection range and selection factor were 126 mm and 4.53 respectively. The size at first maturity of torpedo scad has been reported as 250 mm (TL, unsexed). The mean selection length is generally proportional to the mesh size of the cod end over a certain range. Based on selection factor and length at first maturity for torpedo scad, the optimum mesh size for the trawl cod end for responsible harvesting of the species has been estimated to be 55 mm. However, the cod end mesh size commonly used for fish trawl is 16-30 mm. Use of suitable by catch reduction devices (BRD) can protect the juveniles from exploitation. In trawl nets fitted with oval grid and Big eye BRD, L50 value of M *cordyla* was less than its length at first maturity indicating good exclusion opportunity.

Selectivity parameters *of Megalaspis cordyla*

Selectivity parameters	Value
L25 (TL.mm)	231
L50 (TL,mm)	294

L75 (TL,mm)	357
Selection range (mm)	126
Selection factor	4.52
Length at first maturity (TL, mm, based on Reuben et al., 1992)	250
Cod end mesh size required (mm)	55

Trawl cod end selectivity information is required for biological investigations, fish stock assessment, fisheries management and fishing gear design and development. Based on the selectivity experiments, the cod end mesh size that can be recommended to harvest the mature torpedo scad is estimated as 55 mm for conventional diamond mesh cod ends.

11 | Accessories of Trawl Net

Otter doors or otter boards

Otter boards, the stabilizing devices for trawl nets while under tow, function as hydro-dynamically dependent means for holding the trawl mouth open. Many improvements in the designs of otter boards have come into being to meet the demands of factors, like, the type of ground to be trawled, method of fishing, sizing of trawlers and its towing power Their ability to spread the trawl is derived from inter action of external forces, while under tow and the magnitude of these forces depends mainly on their size and shape of the angle of attack, towing speed and density of sea water. The size of otter board selected should be matched to the trawl gear operated, rather than to the vessel's horse power, which in effect means the amount of twine that has gone into the construction of net in square meter, (Ferro, 1981).

Different designs of otter boards have been tried for pelagic and semi-pelagic trawling and due to their stability in operation with higher towing speed and versatility in mid-water operations along with ease in handling, high aspect ratio vertically curved otter boards (Suberkrub, 1959) are considered highly effective. Polyvalent doors perform the dual purpose of "on" and off-bottom fishing and being a combination of oval and cambered doors, they have the efficiency to traverse the hard ground with increased spreading efficiency. An efficient pair of otter boards will have large spreading force (CL- lift coefficient) and a low drag force (CD- drag coefficient). The efficiency of a pair of otter boards in graphical form is presented in terms of CL/CD. The higher the value, the otter board is considered more efficient by way of achieving a reduced drag force leading to sizeable reduction in fuel consumption. On an average 25% of the total drag of a gear system and 16% of the total fuel consumption are due to resistance offered by the otter boards during trawling operations. Both polyvalent and high aspect ratio Suberkrub otter boards are best suited for semi-pelagic trawling and are capable of achieving a 15% drag reduction and a minimum of 3 to 5 percent saving in fuel consumption.

Designing trawl gear and accessories capable of lowering the towing resistance was not given due attention in the past when fuel prices were low and fish prices rather high. The situation has changed completely and modified designs which will maintain the catch potential but reduce fuel cost have become an absolute necessity at present.

Polyvalent and Suberkrub types of otter boards designed and selected for the trial are cambered doors which offer significantly greater spreading force for a given projected area when compared with flat or "V" form doors having identical board area. The use of camber can also produce improved water flow around the otter board which is considered instrumental to reducing drag and hence the potential for imprpved functioning. A camber of about 10% is usually considered to be a good compromise. Polyvalent doors (1407 x 982 mm) (Fig. 100) tried had a camber of 9%, being the maximum depth of curvature of the main plate expressed as a percentage to the breadth of the board. Providing a slot in polyvalent doors allows smooth water flow around the doors and improve their hydrodynamic performance. For this purpose two slot one each on the upper and lower middle sections were provided in the experimental polyvalent boards.

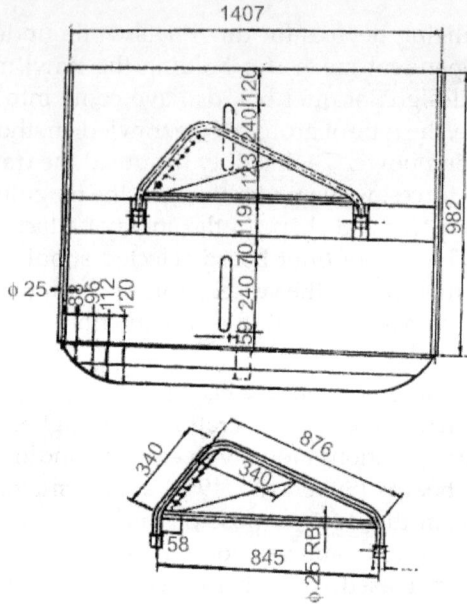

Fig. 100: 1407 x 982 mm Polyvalent otter board

Aspect ratio (H/L) in the case of Suberkrub doors (1350 x 1000 mm) (Fig. 101) tried was 1.35 and when the ratio is progressively increased from the conventional 0.5, the efficiency is improved, but instability at lower towing speed can occur.

For this reason Suberkrub doors are found suitable for off bottom as well as high speed demersal trawling where a higher towing speed is found essential. Mukundan and Hameael (1995) observed the high aspect ratio Suberkrub doors as efficient tools for inshore trawling with less of a drag force and corresponding increase in shearing force in comparison to horizontally curved low aspect ratio doors.

The area and size of the experimental boards were selected matching to the 18 m RMT 8P semi-pelagic trawls, calculated twine area and the weight of the board restricted to the level as calculated from the formula;

W 75 x B two-third = FERRO (op.cit), where W = weight in kg

B = board area in square meter.

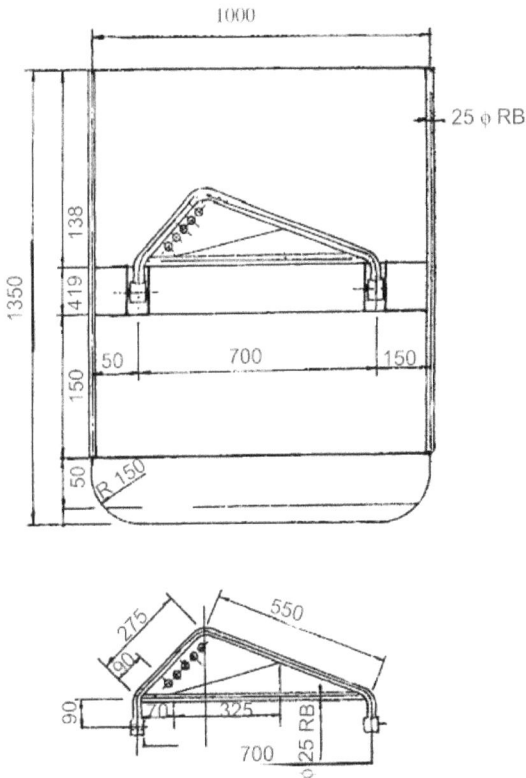

Fig. 101: 1350 x 1000 mm High-Aspect ratio otter board

During the investigations, the two trial of otter boards were operated giving equal chances and keeping the various parameters, such as, fishing grounds, gear, length of bridles, depth of operation, depth to warp ratio, trawling speed and duration of tow constant. Sixty four comparative hauls of one hour duration each were made with both pair of otter boards during 54 fishing days, from a 17.5 mLOA, 277 HP vessel capable of developing a trawling speed range of 2.4 to 3.7 knots while towing against and with the current at 700 rpm.

Higher aspect ratio of Suberkrub doors allow a large spreading force and polyvalent doors are cambered to influence the hydro-dynamic spreading efficiency per unit area, suitable for use in bottom as well as off bottom trawling practices.

The percentage of quality fish caught was 38.4 while operating polyvalent doors, whereas it was 40% when the gear system was rigged with high aspect ratio Suberkrub doors. 8.7% of the total fish caught when the gear was attached to polyvalent doors consisted *ofPampus* sp, which is a high quality catch component and the same was 5% when rigged with Suberkrub doors. The gear when rigged with both the designs of otter boards could develop a vertical opening in the range of 3,8 to 4.5 m well over 20% of the head rope length of the experimental gear which is considered to be the optimum, From the results it could be concluded that both designs of otter boards experimented were efficient accessories for the capture of quality as well as other target species of fish in combination with 18 m semi-pelagic trawl for off bottom trawling in inshore waters.

Otter doors or otter boards form one of the important fishing gear accessories and are in regular use in all trawling operations. Otter boards regulate the mouth opening of the trawl net and as such successful trawling operations to a great extent, depend on their size, weight, shape and behavior under actual tow. The doors are usually made of calculated weights either of hard wood with iron frame and fittings or fabricated entirely out of mild steel plates and rods. While wooden doors suffer heavy damages and natural deterioration due to organic decay, steel doors rapidly wear out due to sea water corrosion. Deteriorating steel plates not only loose mechanical strength, but also get perforated as a result of corrosion and become unfit for any further use, unless plates are renewed from time to time. The initial investment of steel otter doors are considerably high, so also their maintenance and subsequent repairs. A pair of otter boards made of 3 mm thick mild steel plates is estimated to last under constant use, two fishing seasons of eight months each. Considering the present scarcity and restricted supply of mild steel, the high initial investment of steel otter door and their high rates of corrosion under use in a tropical marine environment (6 mils and above per year), the reconditioning the corroded steel otter doors with layers of fiber glass reinforced plastics (FRP) was considered.

Such a procedure adopted and tried under prototype studies has resulted in considerable savings and salvaging of unserviceable steel doors. The versatility of fiber glass scathing in protecting the wooden hulls of fishing boats is well known now. The following are the specifications for the materials required in connection with the seathing operation of steel otter doors.

1. Glass fiber = F 100 Fiberglass (A type) chopped strand mat, 450 g/sq.m
2. Thermosetting resin = Isophthalic polyester resin or the general purpose resin.
3. Accelerator = Cobalt naphthenate or cobalt octoate.
4. Catalyst = Methyl ethyl ketone peroxide.

    Acetone for cleansing purposes, paint brushes, plastic containers, pair of tailoring scissors and metal rollers are the other ancillary materials essential during the work (FRP seathing involves very delicate handling of chemicals and as such necessary precautions have to be taken strictly as per manufacturer's directions).

Before actually starting the seathing work, the total area and shape of the surface have to be detailed out and the chopped strand mat of fiber glass has to be tailored and cut to shape. Polyester resin amounting to about two times the weight of the glass mat may be necessary, which in turn has to be activated by adding 1 % of cobalt naphthenate or cobalt octoate and 1 % methyl ethyl ketone peroxide which will set to gel within 45 minutes at 30 degree Celsius and relative humidity 76%. For easy and convenient handling mixing of 1 kg lot of resin at a time is recommended which will facilitate working with it for nearly 45 minutes at a stretch without undergoing gelation.

Hand lay-up process was adopted for the seathing job throughout in the following manner.

(1) Remove all mil scale and rust either by sand blasting, mild chipping or wire brushing. If perforated, beat up, and level surface.

(2) Remove oil, grease, moisture if any, by thorough scrubbing and cleaning with suitable solvents. Keep the surface dry and clean. The proper adhesion of FRP to the metallic surface depends entirely on the surface preparation.

(3) Perforated areas on the board have to be patched up from both sides using resin mat resin combination before full seaming is attempted. This may be done by fixing resin welted faces of glass fiber mat together sandwiching the corroded steel plate in between. Allow the patch work to dry.

(4) The entire area to be seathed may be fully welted with a generous coat of activated isophthalic resin or a general purpose gel coat. When the coating is still tacky, lay up the first layer of fiber glass chopped strand mat or fiber glass woven roving that had been cut to size already. Use some more activated resin and with a paint brush and metal roller press the seathing on the steel surface for obtaining uniform adhesion. All the comers and crevices and bends have to be worked with glass fiber and resin. No metal surface should be left without seathing. Avoid air bubbles and void spaces on the seathing. If they are there, break them up and rebuild those areas and press it down with a roller. When the first layer is tacky, but not fully dried up, build up the second layer using the mat and resin. If any color scheme is needed, special quality pre-dispersed color pigments can be put to use with the finishing top coat. After two weeks of post-curing, the boards can be commissioned for use.

FRP seathing not only provides extra strength to the corroded plates, but gives them a further lease of active life. The external seathing is tough, rigid and impervious to the effect of salt water and is unaffected either under constant or intermittent immersion or alternate wetting and drying. The seathing provides adequate impact resistance, but when dragged over coarse sand and hard rocky beds, sign of gradual wearing is possible due to surface abrasion. As long as inner steel core is not exposed to sea water or external damages, the board could be put to further use. In case of de-lamination or surface damages, it is enough that seathing is renewed only in the affected areas. Using two layers of FRP chopped strand mat as specified above, protection to such a seathing assure many years of trouble free rough and tough use out of condemned steel otter boards.

Otter boards of different shapes

The successful function of trawling gear depends on the type, size, shape and bridle attachment of the otter boards, which are the main net mouth opening devices. Flat rectangular boards having length greater than breadth are universally employed. Studies to improve the otter boards to suit different boats and fishing conditions have been made by different workers.

Comparative efficiency between three different shaped otter boards, namely, horizontal curved, vertical curved and "V" shaped were employed. Vertical boards are based on the design of Suberkrub (1959); while that of "V"-shaped ones are based on the design of Captain Loo Chi Ho of Taiwan as reported by Dick Brett (1962). The three types of otter boards were operated in rotation with an 18.26 m two seam trawl net having 20 m single sweep wire system on each side. Fishing operations were carried out from 12.76 m wooden boat with 60 HP engine Off Kakinada, Andhra Pradesh during 1970 and 1971.

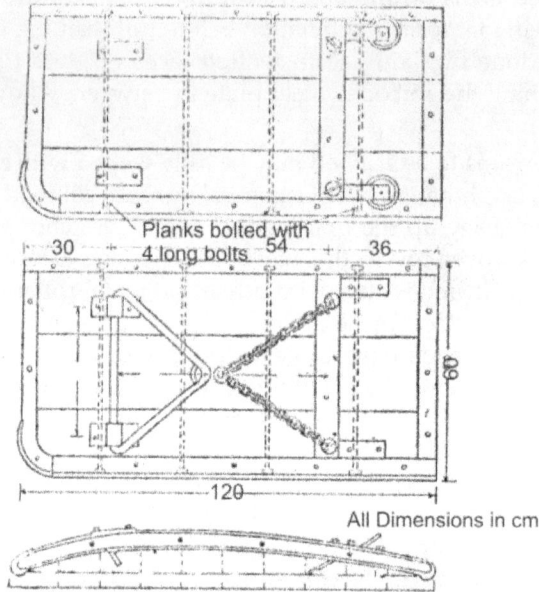

Fig. 102: Design details of horizontal curved otter board (120 x 60 cm)

Specification of otter boards

Type	Horizontal	Vertical curved	"V"-shaped
Material	Wood and iron	Wood and iron	Iron
Length (cm)	120	121	122
Breadth (cm)	60	65	74
Area (sq. m)	0.770	0.780	0.835
Weight in air (kg)	43.5	48.5	61.0
Weight in water (kg)	15.50	21.00	52.00
Angle of attachment of bridle	30 degree	28 degree	40-57 degree
Details of fishing with three types of otter boards			
Type of boardes	Horizontal	Vertical curved	"V"-shaped
Depth of operation (m)	15-45	15-45	15-45
Warp length (m)	90-220	90-220	90-220
Number of hauls	65	65	65
Towing time (hours)	65	65	65

Average towing speed (knots)2	2	2
Average spread between boards (m) 21.63	22.74	20.99
Average warp tension (kg)454.72	459.62	468.92
Cateh hauI hour (kg) 50.45	42.98	52.85

The horizontal curved and "V"-shaped steel otter boards found to be effective, were further studied at different scope ratios. These boards were operated at 15 m, 25 m and 35m depth with scope ratios (Depth : Wire rope length) of 1:4, 1:5 and 1.6 keeping the fishing conditions constant and trawling speed at 2 knots.

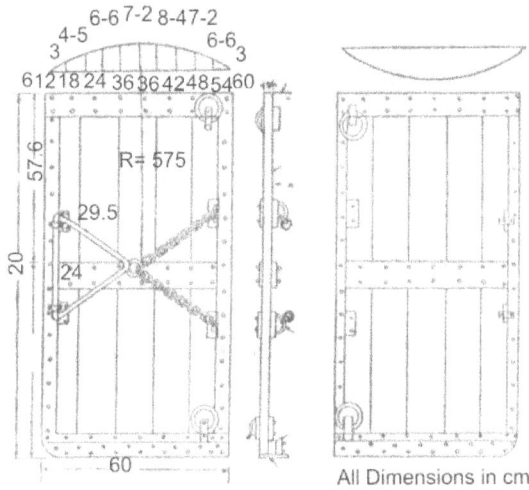

Fig. 103: Design details of vertical curved otter board (120 x 60 cm)
Catch per hour (kg) of different otter boards at different scope ratios

Depth (m)		15			25			35	
Scope ratio	1:4	1:5	1:6	1:4	1:5	1:6	1:4	1:5	1:6
Horizontal curved	7.0	26.5	39.0	12.0	29.0	36.6	13.0	37.0	92.0
"V"-shaped	16.0	14.0	10.1	48.5	59.3	46.3	64.8	100.0	84.0

It is seen that out of three scope ratios, the horizontal curved board has the maximum scope ratio 1:6 and that for "V"-shaped 1:5 with an exception at 15 m depth.

The net operated with "V-shaped otter board caught more fish when compared to nets fitted with horizontal and vertical curved board had the maximum scope ratio of 1:6 and for "V"-shaped 1:5 with an exception at 15 m depth. "V"-shaped otter board caught more fish when compared to nets fitted with horizontal and vertical curved boards by 4.64 and 18.67 percent respectively.

The percentage composition of catch reveals that there is no appreciable difference in the catch composition. But the percentages of off bottom fishes are slightly more in the net with vertical curved boards. This suggests its use for both bottom as well as off bottom fishes, with minor variation in fishing techniques even though these boards are intended for mid-water trawling.

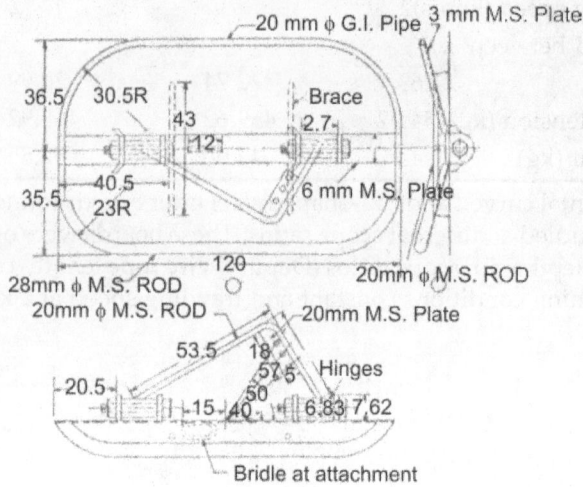

Fig. 104: Design details of "V"-shaped otter board

The average horizontal spread of the trawl mouth shows that vertical curved otter boards developed more mouth opening horizontally, but with less catch rate. So a further attempt was made to correlate the catch rate with horizontal opening of the net in each case. The percentage frequency of spread shows the corresponding catch per trawling hour of "V"-shaped board gave more frequency in the spread range of 20.1 to 21.5 m which in turn gave the highest catch rate. Similarly the horizontal curved boards gave maximum frequency of horizontal spread at the ranges of 21.6 to 23.0 m which gave the catch rate next to "V"-shaped boards.

Horizontal spread of each otter board at various depths between 15 to 50 m together with corresponding catch rate and the warp tension show that good catches were obtained when the gears were operated between 30 to 40 m depths. The horizontal spread was also more at these depths. Within these depth ranges, the horizontal spread and catch of gear with "V"-shaped doors were more, followed by rectangular curved doors. Comparatively higher warp tension at lower depths may be due to bottom friction, which in turn reduces the horizontal spread. At higher depths the tension appears to have been stabilized at 455, 450 and 475 kg for each of the otter boards. The variation in tension between gears is significant and their averages worked out to be 454.72, 459.62 and 468.92 kg respectively. It is also seen that warp tension between horizontal and vertical curved boards is not significant, whereas warp tension between horizontal curved and "V"-shaped boards is significant.

12

Sieve trawl – By catch Reduction Device

The term by catch means, that portion of the catch other than target species caught while fishing, which are either retained or discarded. Average annual global discards has been estimated to be about 7.3 million tonnes based on weighted discard rate of 8% during 1992-2004 period. Trawl fisheries for shrimp and demersal fin fish account for over 50% of the total estimated global discards. (Kelleher, 2004)

In India, the by catch problem is acute due to the multi-species nature of fisheries. Kelleher (2004) has estimated total by catch discards in Indian fisheries at 57917 tonnes which formed 2.03% of total landings. The dominant varieties among the discards were fin fishes, crabs and stomatopods.

Sieve net is a large mesh funnel fixed inside the net, which guides the fish to a second cod end with large diamond mesh netting, while shrimps pass through large meshes accumulate in the main cod end. Sieve net (also known as veil net) with out outlet cod end is made mandatory under EU legislation in European brown shrimp fisheries. Sieve net is used in commercial shrimp fleets of the Netherlands, U.K, France, Germany and Belgium. Sieve net was found to be the most effective trawl modification, which reduce discard level of juvenile fish and shrimps and was recommended for mandatory use in beam trawls in U.K.

In Indian Oceans experimental trials were carried out during September to December 2006 and March, 2007 from a 17.5 m LOA trawler with 277 HP and a 15.24 m LOA trawler with 223 HP engine. The experimental fishing operations were conducted during day time at depths ranging between 9 and 32 m in the seas off Cochin. The sieve net was attached to the body of a commercial shrimp trawl of 32.4 m rigged with "V" type steel otter boards of 1420 x 790 mm size (80 kg each) and 20 m double bridles. The net was made of knotted polyethylene (PE) netting with mesh size of 50 mm in the front part and decreasing to 30 mm in the aft part of the net. The cod end made of knotted PE netting of 20 mm mesh size was provided with a protective cover made of 120 mm mesh size and 3 mm diameter PE netting.

Sieve net design

Designs of by catch reduction device (BRD) included net (i) with 60 mm diamond mesh funnel inside the net and 80 mm diamond mesh outlet cod end (Sieve net-60) and (ii) with 50 mm diamond mesh funnel inside the net and 60 mm diamond mesh outlet cod end (Sieve net-50) (Fig. 105).

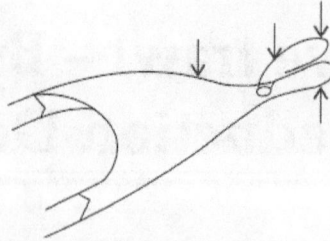

Fig. 105: Perspective view of Sieve net BRD installed in the trawl net

In Sieve net 60, a funnel made of 60 mm meshes netting (135 meshes in circumference in the leading edge, 19 meshes in circumference in the hind edge and 70 meshes in depth) was used for separation of shrimps. The hind end of the tunnel is opening to an outlet cod end with 80 mm mesh size of 4 m length and 60 meshes in circumference. In Sieve net 50, a 50 mm mesh funnel (162 meshes in circumference in the leading edge, 22 meshes in circumference in the hind edge and 84 meshes in depth) was used. The hind end of the funnel opened to an outlet cod end of 4 m length and 70 meshes in circumference fabricated out of 60 mm mesh netting. The outlet cod end of the experimental sieve nets was provided with small mesh (12 mm) cover, which was 2.5 times the dimensions of the outlet cod end in order to retain the excluded catch.

In the 60 mm outlet cod end four species, namely, *Charybdis lucifera, Charybdis feriatus, Carangoides armatus and llpeneus suiphurus* were fully retained. Seven species, namely, *Portunus pelagicus, Portunus sanguinolentus, Mene maculate, Leiognathus brevirostris, Nemipterus japonicus, Saurida undosquamis* and *Dolcia ovis* showed more than 50% exclusion and out of total 50 species encountered in outlet cod end, 39 species showed retention up to 50%.

In sieve net 50 length wise retention and exclusion characteristics in respect of four species, namely, *Lepturacanthus savala, Megalaspis cordyla, Pampus argenteus* and *Otolithes rubber* are given in Fig. 106.

Length class of 101=120 mm of *Lepturacanthus savala* showed 100% exclusion, 141-160 mm length class showed full retention and length classes from 161-280 mm showed retention in the range of 43 to 96%.. *Megalaspis cordyla* in the length class of 51-100 mm was fully excluded, and there was an increasing trend in the exclusion of length classes from 101 to 300 mm. Length classes *of Otolithes rubber* from 51 to 170 mm showed an increasing trend in exclusion rate from 10 to 70% and length classes from 170 to 250 mm were fully excluded. *Pampas argenteus* in the length class of 21 to 30 mm was fully retained, length classes of 31 to 100 mm showed an increasing trend in exclusion rate in the range of 41 to 98% and length classes from 101 to 130 mm showed 100% exclusion.

Sieve net was found to be an effective design for the reduction of fish by catch in different waters of the world. The commercial version of Sieve net design (the funnel having a mesh size of 70 mm and an outlet cod end with mesh size of 80 mm) provide escapement opportunity for juveniles, small fishes and invertebrates. Using Sieve net in Belgium fishery has shown by catch exclusion rates of 29-50% in different seasons with less than 15% loss of shrimps. It is less effective in saving fishes less than 10 cm. Four designs of Sieve net were evaluated in commercial shrimp beam trawling. During the evaluation, sieve net was found to be the most effective trawl modification which reduce discard levels of juvenile fish and shrimps. Sieve net reduced small shrimp *(Crangon crangon)* to the tune of 29% by weight was recommended mandatory use in beam trawls in U.K. Hoist (2004) CEFAS (2003) reported the use of a cone shaped large mesh netting with bottom opening known as Veil net, which is similar in operational principle to Sieve net. This device reduced the retention of juvenile fish and invertebrates in the trawls and recommended this technology for use in other fisheries.

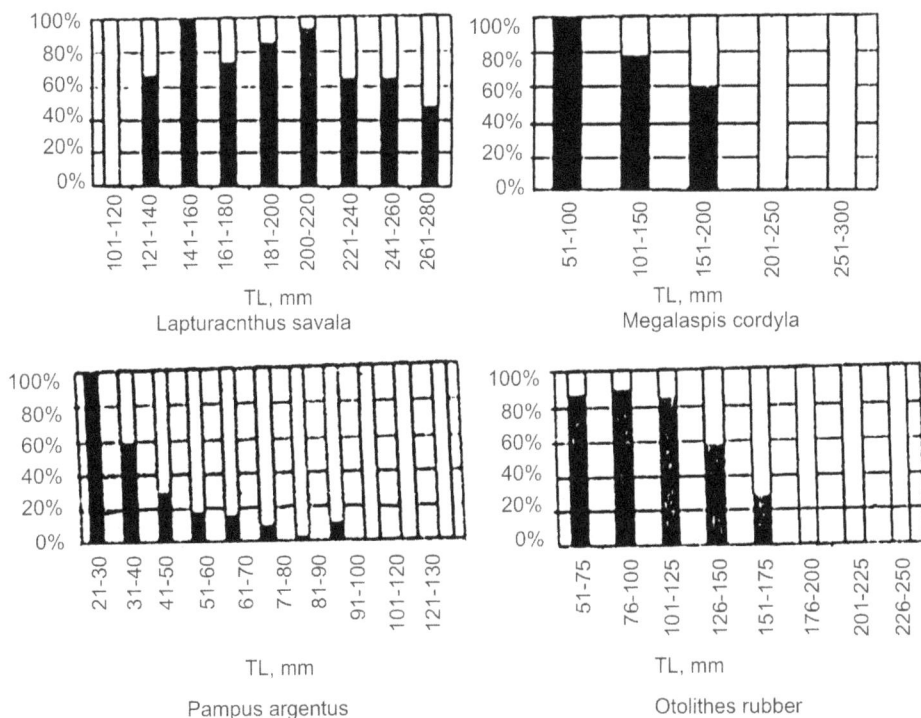

Lapturacnthus savala

Megalaspis cordyla

Pampus argentus

Otolithes rubber

Fig. 106 - Lengthwise retention and exclusion
of selected species from Sieve net 50

Complete exclusion of by – catch from shrimp trawls may not be always acceptable to the fishermen, as a part of the by catch constituted by large marketable species, often contribute to the profitability of trawl operations in the tropical fisheries. Sieve net designs which are appropriately adapted to regional fisheries in terms of mesh sizes of the outlet and main cod ends is expected to be acceptable and could lead to significant reduction in the mortality of juveniles during shrimp trawling. Sieve net 50 (50 mm diamond mesh funnel and outlet cod end of 60 mm mesh size) has functioned poorly in terms of target catch retention which was only about 80%, making this design unacceptable for commercial use. Sieve net 60 with 60 mm diamond mesh funnel inside the net and 80 mm diamond mesh outlet cod end has been able to exclude substantial quantities of by catch including juveniles while keeping shrimp loss at about 4.5% and retaining larger marketable by catch species. In addition it is also possible to adapt the sieve net to retain shrimp catch and efficiently exclude jelly fish when they abound in the shrimp fishing grounds by keeping the outlet cod end open. Trawl fishermen can thus reduce the sorting time on board resulting in an increase in the useful fishing time and can enhance the profitability of trawl operation. Sieve net 60 has potential for adoption in tropical trawl fisheries, in order to minimize the impact of shrimp trawling on juveniles and non-targeted by catch species.

By catch characteristics of shrimp trawls

By catch taken by the shrimp trawl fishery is an important issue in the management of fisheries resources. Given the perceived high mortality of the different fish stocks other than shrimps in tropical countries like India by catch issue is more complex due to the multi-species and multi-gear nature of the fisheries. The changing perspective of by catch itself offers the greatest challenge is yesterday's by catch becomes today's target catch. Quantum of by catch landed or discarded may depend on factors affecting selectivity of trawls, such as, cod end mesh size, mesh sizes of the wings and belly sections, vertical opening of the trawl mouths, ground rope rigging and bottom contact, overall length of the trawl, otter boards, and bridle arrangements, speed and duration of tow, trip duration (single day or multi-day fishing) storage and preservation facilities available on board, variation in seasonal abundance of by catch species and juveniles and variation in export and domestic market demands for target and by catch species.

The diversity of species is the main cause of the higher magnitude of discards found in tropical waters. With the decline of the shrimp catch, the by catch began to contribute significantly to the overall income of the shrimp trawlers. It is significant to note that among by catch about 40% consisted of juveniles and those in the early stages of development which are invariably discarded leading to the depletion of the resources.

A shrimp trawl of 29 rn head rope with 20 mm diamond mesh cod end, rigged with V-type steel otter boards of 1420 x 790 mm (80 kg each) and 20 m double bridles were used for experimental fishing in coastal waters off Cochin during April 2004 to December, 2006 with a total of 690 hauls with trawling duration varying from 0.75 to 2.0 hours. The catch from individual hauls was examined separately and was categorized into target catch (shrimps) and non-targeted catch or by catch which included all species other than shrimps and weight of each species was recorded to the nearest gram.

Two hundred eighty one species have been encountered in the trawl catch, off south west coast of India. The catch included 191 species of fishes, 11 species of shrimps, 3 species of lobsters, 13 species of crabs, 11 species of cephalopod, 44 species of mollusks shells, 2 species of echinoderms, 2 species of jelly fishes, 2 species of stromatopod and one species of sea snake and one species of sea turtle. One hundred ninty one species of fishes belonged to 12 orders and 59 families and 109 genera. Eleven shrimp species belonging to 4 families and 13 crab species belonging to 5 families have been identified. Eleven cephalopod species belonged to 3 orders and 3 families. Molluscan species belonged to 22 families and jelly fishes belonged to 2 families.

List of species occurring in trawl by catch off Cochin

Finfishes

Order: Rajiformes;

Family: Dasyatidae

1. *Dasyatis kuhlii* (Muller & Henle, 1841)
2. *Himantura bleekeri* (Blyth, 1860)
3. *Himantura uarnak* (Forsskal, 1775)
4. *Himantura gerrardi* (Gray, 1851)

Family : Myliobatidae

5. *Aetobatus narinari* (Euphrasen, 1790)

Order: Carcharhiniformes

Family : Carcharhinidae

6. *Rhizoprionodon **acutus*** (**Ruppell,** 1837)
7. *Scoliodon laticaudus* (Muller & Henle, 1838)

Family: Sphyrnidae

8. *Eusphyra blochii* (Cuvier, 1816)
9. *Sphyrna zygaena* f Linnaeus, 1758)

Order : Anguilliformes Family : Congridae

10. *Uroconger lepturus* (Richardson, 1845)

Family : Ophichthidae

11. *Pisodonophis cancrivorus* (Richardson, 1845)
12. *Leiuranus semicinctus* (Lay & Bennett, 1839)
13. *Lamnostoma orientalis* (McClelland, 1844)

Family : Muraenesocidae

14. *Congresox talabonoides* (Bleeker, 1853)

Order : Clupeiformes

Family : Chirocentridae

15. *Chirocentrus dorab* (Forsskal, 1775)
16. *Chirocentrus nudus* (Swainson, 1 839)

Family : Ctupeidae

17. *Anodontostoma chacunda* (Hamilton, 1822)
18. *Dussumieria acuta* (Valenciennes, 1 847)
19. *Escualosa thoracata* (Valenciennes, 1847)
20. *Opisthopterus tardoore* (Cuvier, 1829)
21. *Sardinella albella* (Valenciennes, 1 847).
22. *Sardinella fimbriata* (Valenciennes, 1847)
23. *Sardinella gibbosa* (Bleeker, 1849)
24. *Sardinella longiceps* (Valenciennes, 1847)

Family : Pristigasteridae

25. *Ilisha elongate* (Anonymous (Bennett, 1830)
26. *Ilisha filigera* (Valenciennes, 1 847)
27. *Pellona ditchella* (Valenciennes, 1847)

Family : Engraulidae

28. *Encrasicholina devisi* (Whitley, 1940)
29. *Encrasicholina heteroloba* (Ruppell, 1837)
30. *Encrasicholina punctifer* (Fowler, 1938)
31. *Stolephorus commersonnii* (Lacepede, 1803)
32. *Stolephorus indicus* (van Hasselt, 1823)
33. *Stolephorus insularis* (Hardenberg, 1933)
34. *Stolephorus waitei* (Jordan & Seale, 1926)
35. *Thryssa dussttmieri* (Valenciennes)
36. *Thryssa kammalensis* (Bleeker, 1849)
37. *Thryssa malabarica* (Bloch, 1795)
38. Thryssa mystax (Bloch & Schneider, 1801)
39. *Thryssa purava (Hamilton, 1822)*
40. *Thryssa setirostris* (Broussonet, 1782)

Order Siluriformes

Family: Ariidae

41. *Arius arius* (Hantiltom, 1822)
42. *Arius jelia* (Day, 1877)

43. *Arius sona* (Hamilton, 1822)
44. *Arius maculatus* (Thunberg, 1792)
45. *Nemapteryx caelata* (Valenciennes, 1840)
46. *Arius thalasmus* (Ruppell, 1837)

Family: Plotosidae

47. *Plotosus lineatus* (Thunberg, 1787)

Family: Synodontidae

48. *Saurida undosquamis* (Richardson, 1848)
49. *Saurida tumbil* (Bloch, 1795)

Order: Syngnathiformes

Family: Fistularidae

50. *Fistularia petimba* (Lacepede, 1803)

Order: Scorpaeniformes

Family : Scorpaenidae

51. *Pterois volitans* (Linnaeus, 1758)
52. *Pterois russelii* (Bennett, 1831)

Family: Platycephalidae

53. *Platycephalus indicus* (Linnaeus, 1758)
54. *Gramrnoplites scaber* (Lennaeus, 1758)
55. *Thysanophrys celebica* (Bleeker, 1854)
56. *Cociella crocodilus* (Tilesius, 1812)

Family: Dactylopteridae

57. *Dactyloptera macracantha* (Bleeker, 1854)

Family: Synanceiidae

58. *Minous monodactylus* (Bloch & Schneider, 1801)
59. *Minous dempsterae* (Eschmeyer, Hallacher & Rama-Rao, 1979)
60. *Synanceia horrida* (Linnaeus, 1766)
61. *Leptosynanceia asteroblepa* (Richardson, 1844)

Order: Beryciformes

Family : Holocentridae

62. *Myripristis adusta (Bleeker, 1853)*

Order: Perciformes Family: Teraponidae

63. *Terapon jarbua* (Forsskal, 1775)
64. *Terapon theraps* (Cuvier, 1829)
65. *Terapon puta* (Cuvier, 1829)
66. *Pelates quadrilineatus* (Bloch, 1790)

Family : Serranidae

67. *Epinephelus latifasciatus* (Temminck & Schlegel, 1842)
68. *Epinephelus diacanthus* (Valenciennes, 1828)
69. *Epinephelus merra* (Bloch, 1793)
70. *Epinephelus tauvina* (Forsskal, 1775)
71. *Epinephelus areolatus* (Forsskal, 1775)
72. *Epinephelus chlorostigma* (Valenciennes, 1828)

Family: Priacanthidae

73. *Priacanthus hamrur* (Forsskal, 1775)

Family : Apogonidae

74. *Apogon aureus* (Lacepede, 1802)
75. *Apogon fasciaius* (White, 1790)

Family : Pomacentridae

76. *Neopomacentrus sindensis* (Day, 1873)

Family : Haemulidae

77. *Pomadasys maculatum* (Bloch, 1793)

Family: Lutjanidae

78. *Lutjanus malabaricus* (Bloch & Schneider, 1801)
79. *Pinjalo pinjalo* (Bleeker, 1850)
80. *Lutjanus argentimaculatus* (Forsskal, 1975)
81. *Lutjanus lutjanus* (Bloch, 1790)

Family : Lethrinidae

82. *Lethrinus nebulosus* (Forsskal, 1775)
83. *Lethrinus ornatus* (Valenciennes, 1830)
84. *Lethrinus miniatus* (Bloch & Schneider, 1801)

Family : Nemipteridae

85. *Nemipterus japonicus* (Bloch, 1791)
86. *Nemipterus mesoprion* (Bleeker, 1853)

Family : Gerreidae

87. *Gerres oyena* (Forsskal, 1775)
88. *Gerres filamentosus* (Cuvier, 1829)
89. *Gerres erythrourus*(Bloch, 1791)
90. *Gerres limbatus*(Cuvier, 1830)

Family: Mullidae

91. *Upeneus sulphureus* (Cuvier, 1829)
92. *Upeneus vittatus* (Forsskal, 1775)
93. *Upeneus tragula* (Richardson, 1864)

Family : Sillaginidae

94. *Sillago sihama* (Forsskal, 1775)

Family : Lactariidae

95. *Lactarius lactarius* (Bloch & Schneider, 1801)

Family: Sciaenidae

96. *Johnius amblycephalus* (Bleeker, 1855)
97. *Johnius borneensis* (Bleeker, 1851)
98. *Johnius carouna* (Cuvier, 1830)
99. *Johnius carutta* (Bloch, 1793)
100. *Johnius dussumieri* (Cuvier, 1830)
101. *Kathala axillaris* (Cuvier, 1830)
102. *Nibea maculata* (Bloch & Schneider, 1801)
103. *Otolithes cuvieri* (Trewavas, 1974)
104. *Otolithes rubber* (Bloch & Schneider, 1801)
105. *Otolithoides biauritus* (Cantor, 1849)
106. *Protonibea diacanthus* (Lacepede, 1802)
107. *Daysciaena albida* (Cuvier, 1830)

Family: Leiogoathidae

108 *Gazza minute,* (Bloch, 1795)
109. *Leiognathus bindus* (Valenciennes, 1835)
110. *Leiognathus brevirostris* (Valenciennes, 1835)
111. *Leiognathus daura*(Cuvier, 1829)
112. *Leiognathus dussumieri* (Valenciennes, 1835)
113. *Leiognathus elongatus* (Gunther, 1874)
114. *Leiognathus equulus* (Forsskal, 1775)
115. *Leiognathus splendens* (Cuvier, 1829)

116.   *Secutor insidiator* (Bloch, 1787)
117.   *Secutor ruconius* (Hamilton, 1822)

Family : Carangidae

118.   *Aleetis ciliaris* (Bloch, 1793)
119.   *Aleetis mdicus* (Ruppell, 1830)
120.   *Alepes djedaba* (Forsskal, 1775)
121.   *Alepes kleini* (Bloch, 1793)
122.   *Atropus atropus* (Bloch & Schneider, 1801)
123.   *Atule mate* (Cuvier, 1833)
124.   *Carangoides armatus* (Ruppell, 1830)
125.   *Carangoides malabaricus* (Bloch & Schneider, 1801)
126.   *Carangoides oblongus* (Cuvier, 1833)
127.   *Carangoides praeustus* (Anonymous (Bennett), 1830)
128.   *Caranx ignobilis* (Forsskal, 1775)
129.   *Caranx sexfasciatus* (Quoy & Gaimard, 1825)
130.   *Decapterus russelli* (Ruppell, 1830)
131.   *Gnathanodon speciosus* (Forsskal, 1775)
132.   *Megalaspis cordyla* (Linnaeus, 1758)
133.   *Parastromateus niger* (Bloch, 1795)
134.   *Scomberoides lysan* (Forsskal, 1775)
135.   *Scomberoides tala* (Cuvier, 1832)
136.   *Scomberoides tol* (Cuvier, 1832)
137.   *Selar crumenophthalmus* (Bloch, 1793)
138.   *Trachinotus blochii* (Lacepede, 1801)
139.   *Uraspis uraspis*(Gunther 1860)

Family : Polynemidae

140.   *Eleutheronema tetradactylum* (Shaw, 1804)
141.   *Filimanus heptadactyla* (Cuvier, 1829)
142.   *Filimanus similes* (Feltes, 1991)
143.   *Leptomelanosoma indicum* (Shaw, 1804)

Family : Sphyraenidae

144.   *Sphyraena barracuda* (Walbaum, 1792)
145.   *Sphyraena forsteri* (Cuvier, 1829)
146.   *Sphyraena jella*(Cuvier, 1829)
147.   *Sphyraena obtusata* (Cuvier, 1829)

Family : Gobiidae

148. *Oxyurichthys paulae* (Pezold, 1998)
149. *Trypauchen vagina* (Bloch & Schneider, 1801)

Family: Trichiuridae

150. *Lepturacanthus savala* (Cuvier, 1829)
151. *Trichiurus lepturus* (Linnaeus, 1758)

Family: Stromateidae

152. *Pampas argenteus* (Euphrasen, 1788)
153. *Pampus chinensis* (Euphrasen, 1788)

Family : Ambassidae

154. *Ambassis ambassis* (Lacepede, 1802)
155. *Ambassis commersonnii* (Cuvier, 1828)
156. *Ambassis gymnocephalus* (Lacepede, 1802)

Family : Mugilidae

157. *Liza microlepis* (Smith, 1846)
158. *Liza parsia* (Hamilton, 1822)
159. *Liza subviridis* (Valenciennes, 1835)
160. *Liza tade* (Forsskal,1775)
161. *Mugil cephalus* (Linnaeus, 1758)
162. *Valamugil cunnesius* (Valenciennes, 1836)
163. *Valamugil speigleri* (Sleeker, 1858-59)

Family: Menidae

164. *Mene maculata* (Bloch & Schneider, 1801)

Family: Scatophagidae

165. *Scatophagus argus* (Linnaeus, 1766)

Family : Scombridae

166. *Rastrelliger kanagurta* (Cuvier, 1816)
167. *Scomberomorus commerson* (Lacepede, 1899)
168. *Scomberomorus guttatus* (Bloch & Schneider, 1801)
169. *Scomberomorus lineolatus* (Cuvier, 1829)

Family : Siganidae

170. *Siganus canaliculatus* (Park, 1797)
171. *Siganus javus* (Linnaeus, 1766)

Family : Acanthuridae

172.   *Acanthurus mata* (Cuvier, 1829)

Family : Uranoscopidae

173.   *Uranoscopus marmoratus (* Cuvier, 1829)

Family : Drepaneidae

174.   *Drepane punctata (* Linnaeus, 1758)

Family : Pempheridae

175.   *Pempheris mangula* (Cuvier, 1829)
176.   *Pempheris oualensis* (Cuvier, 1831)

Order : Pleuronectiformes

Family : Samaridae

177.   *Samaris cristatus (* Gray, 1931)

Order: Beloniformes Family: Hemirhamphidae

178.   *Rhynchorhamphus georgii (* Valenciennes)

Order: Pleuronectiformes

Family: Cynoglossidae

179.   *Cynoglossus are!* (Schneider, 18010
180.   *Cynoglossus bilineatus* (Lacepede, 18020)
181.   *Cynoglossus macrostornus* (Norman, 1928)
182.   *Cynoglossus dubius (* Day, 1873)

Family : Soleidae

183.   *Zebrias quagga* (Kaup, 1858)

Family: Paralichthyidae

184.   *Pseudorhombus arsius* (Hamilton, 1822)

Order: Tetraodontiformes

Family: Triacanthidae

185.   *Triacanthus biaculeatus (* Bloch, 1786)
186.   *Triacanthus nieuhofii* (Bleeker, 1852)
187.   *Pseudotriacanthus strigilifer* (Cantor, 1849)

Family : Diodontidae

188.   *Cyclichthys orbicularis* (Bloch, 1785)

Family : Tetraodontidae

189.   *Lagocephalus spadiceus (* Richardson, 1845)
190.   *Lagocephalus inermis (* Temminck & Schlegel, 1850)
191.   *Chelonodon patoca (* Hamilton, 1822)

SHRIMPS

Order: Decapoda

Family: Penaeidae

192.   *Fenneropenaeus indicus (* H. Milne Edwards, 1837)
193.   *Metapenaeus affinis(H.* Milne Edwards, 1837)
194.   *Metapenaeus dobsoni (* Miers, 1878)
195.   *Metapenaeus monoceros (* Fabricius, 1798)
196.   *Parapenaeopsis stylifera (* H.Milne Edwards, 1837)
197.   *Penaeus semisulcatus* (De Mann, 1844)
198.   *Penaeus monodon (* Fabricius, 1798)
199.   *Trachypenaeus curvirostris (* Simpson, 1860)

Family: Hippolytidae

200.   *Exhippolysmata ensirostris (* Kemp, 1914)

Family : Sergestidae

201.   *Acetes indicus* (H.Milne Edwards, 1830)

Family : Alphidae

202.   *Alpheus malabaricus (* Fabricius, 1798)

LOBSTERS

Order: Decapoda

Family: Palinuridae

203.   *Palinurus homarus (* Linnaeus, 1758)
204.   *Palinurus ornatus* (Fabricius, 1798)

Family: Scyllaridae

205.   *Thenus orientalis* (Lund, 1793)

CRABS

Order: Decapoda

Family: Lucosidae

206. *Philyra scabriucuila* (Fabricius, 1798)

Family: Calappidae

207. *Calappa lophos* (Herbst, 1782)

Family: Portunidae

208. *Charybdis feriatus* (Linnaeus, 1758)
209. *Charybdis lveifeara* (Fabricius, 1798)
210. *Charybdis natator* (Herbst, 1789)
211. *Podophthalmus vigil* (Fabricius, 1798)
212. *Portunus pelagicus* (Linnaeus, 1766)
213. *Portunus sanguinolentus* (Herbst. 1783)
214. *Scylla serrata* (Forskal, 1775)

Family : Matutidae

215. *Ashtoret lunaris* (Forskal, 1775)
216. *Matura planipes* (Eabricius, 1798)

Family: Epialtidae

217. *Doclea ovis* (Fabricius, 1787)
218. *Doclea rissoni* (Leach, 1815)

STOMATOPODS

Order: Stomatopoda

Family : Squillidae

219. *Oratos quilla nepa* (Latreille 1828)
220. *Squilla* sp.

CEPHALOPODS

Order: Sepiida

Family : Sepiidae

221. *Sepia aculeata* (van Hasselt, 1835)
222. *Sepia pharaonis* (Ehrenberg, 1831)
223. *Sepiella inermis* (van Hasselt, 1835)

Order; Teuthida

Family: Loliginidae

224. *Doryteuthis singhalensis* (Ortmann, 1891)
225. *Uroteuthis duvaucelii* (d'Orbigny, 1835)

Order: Octopoda

Family: Octopodidae

226. *Amphioctopus aegina* (Gray, 1849)
227. *Amphioctopus membranaceus* (Quoy & Gaimard, 1832)
228. *Cistopus indicus* (Rapp, 1835)
229. *Octopus globosus* (Appelof, 1886)
230. *Octopus vulgaris* (Cuvier, 1797)

SHELLS

Order: Arcoida

Family: Arcidae

231. *Anadara (Cunearca) rhombea* (Born, 1780)
232. *Anadara granosa* (Linnaeus, 1758)
233. *Barbatia bistrigata* (Dunker, 1866)
234. *Scapharca inaequivalvis* (Bruiguiere, 1789)
235. *Trisidos tortuosa* (Linnaeus, 1758)

Order: Neogastropoda

Family: Babyloniidae

236. *Babylonia spirata* (Linnaeus, 1758)
237. *Babylonia zeylanica* (Bniguiere, 1789)

Family : Buccinidae

238. *Cantharus spiralis* { Gray, 1839)

Family: Turridae

239. *Lophiotoma indica* (Roding, 1798)
240. *Turricula javana* (Lamarc, 1816)
241. *Turris amicta* (E.A. Smith, 1877)

Family: Harpidae

242.   *Harpa major (* Roding, 1798)

Family: Clavatulidae

243.   *Clavatula virgineus* (Dillwyn, 1817)

Family: Muricidae

244.   *Murex (Murex) carbonnieri (* Jousseaume, 1881)
245.   *Rapana bulbosa (* Solander, 1817)
246.   *Rapana rapiformis (* Born, 1778)

Family : Fasciolariidae

247.   *Fusinus nicobaricus* (Roding, 1798)

Family: Melongenidae

248.   *Hemifusu spugilinus (* Born, 1778)
249.   *Pugilina cochlidium* (Linnaeus, 1758)

Order: Littorinimorpha

Family; Bursidae

250.   *Bufonaria echinata* (Link, 1807)

Family: Ficidae

251.   *Ficus ficus* (Linnaeus, 1758)
252.   *Ficus gracilis* (G.B. Sowerby I1825)

Family : Naticidae

253.   *Glossaulax didyma* (Roding, 1798)
254.   *Natica litneata (* Lamarck, 1838)
255.   *Natica vitellus (* Linnaeus, 1758)

Family: Cassidae

256.   *Phalium canaliculatum (* Bruguiere, 1792)
257.   *Semicassis bisulcata (* Schubert & Wagner, 1829)

Family: Rostellariidae

258.   *Strombus plicatus sibbaldi (* Sowerby, 1842)
259.   *Tibia curta (* G.B. Sowerby II, 1842)

Family: Tonnidae

    260.   *Tona dolium (*Linnaeus, 1758)

Order: Veneroida

Family: Veneridae

    261.   *Dosinia cretacea* (Reeve, 1851)
    262.   *Marcina opima(* Gmelin, 1791)
    263.   *Meretrix casta (*Chemnitz, 1782)
    264.   *Meretrix meretrtx (*Linnaeus, 1758)
    265.   *Paphia malabarica* (Chemnitz, 1782)
    266.   *Paphia textile (Gmelin,* 1791)
    267.   *Sunetta scripta (*Linnaeus, 1758)

Family: Donacidae

    268.   *Donax scortum* (Linnaeus, 1758)

Order: Myoida

Family: Pholadidae

    269.   *Pholas orientalis* (Gmelin, 1791)

Family: Cardiidae

    270.   *Cardium flavum (* Linnaeus, 1758)

Order: Caenogastropoda

Family: Turritellidae

    271.   *Turritella acutangula (* Linnaeus, 1758)
    272.   *Turritella attenuata (*Reeve, 1849)

Order: Archaeogastropoda

Family : Trochidae

    273.   *Umbonium vestiarium (*Linnaeus, 1758)

Order ; Dentaliida

Family ; Dentaliidae

    274.   *Dentalium octangulatum* (Donovan, 1804)

ECHINODERMS

Order: Paxillosida

Family: Astropectinidae

275. *Astropecten* spp.

Order: Clypeasteroida

Family: Lagnidae

276. *Laganum depressum (*Lesson, 1841)

JELLY FISH Order: Rhizostomeae

Family : Catostylidae

277. *Crambionella stuhlmanni* (Chun, 1896)

Order: Semaeostomeae

Family: Ulmaridae

278. *Aurelia solida* (Browne, 1905)

TURTLES

Order: Testudines

Family: Cheloniidae

279. *Lepidochelys olivacea* (Eschscholtz, 1829)

SEA SNAKES

Order: Squamata

Family: Elapidae

280. *Aipysurus laevis (*Lacepede, 1804)

By catch was generated at levels exceeding 10 kg/hour during January-March and August-November and at levels less than 19 kg/hour during April-July and December. Organisms other than fish dominated in the by catch during May, August and September, while fishes dominated during other months. Shrimps of marketable size accounted for a small percentage of total trawl landings. The rest of the catch consisted of by catch consisting of variety of fishes, cnidarians, mollusks, crustaceans, echinoderms and juveniles of fish which fetch relatively low value.

By catch formed 79.18% (35902 tonnes) of total shrimp trawl landings in India, which was utilized either for human consumption or as fish meal and fish manure. The by catch discards from mechanized trawlers operating in Indian EEZ was estimated at 1.2 million tones.

Various types of by catch reduction technologies have been developed in the fishing industry around the world in order to improve the selectivity of the shrimp trawls and minimize the impact of trawling on non-target resources and juveniles. The degree of adoption of by catch reduction technologies is strongly dependent on robustness of the fisheries management system. By catch reduction technologies have been mandated and effectively implemented in several scientifically managed fisheries in the world. But its adoption in less effectively managed fisheries require the active involvement of stake holders in the process, supported by a system of incentives, disincentives, education and training.

By catch reduction device (BRDs)

By catch reduction device (BRDs) have been developed taking into consideration the differential behavior patterns, such as, differences in swimming speed and vertical distribution and size selectivity of targeted and non-targeted organisms. The fish generally are active and are capable of swimming against the water flow inside the net and could escape if an opportunity is provided, while the shrimps are carried by the flow of water into the cod end (Fig. 105).

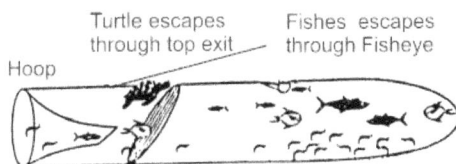

Fig. 107 : Schematic diagram of by catch reduction devices

Most of the BRDs have been developed through intensive research, taking into account the characteristics of the fishery and geographical peculiarities of the region. A classification of BRDs based on the structure, materials used and principles of operation is given below.

Soft BRDs – – – – – – –Hard BRDs – – – – – –> Combination BRDs

<– – – – – – – – – – – –I– – – – – – – – – – –>

Soft BRDs	Hard BRDs	Combination BRDs
BRDs with grids	BRDs with slots	Semi-flexible BRDs
Flat grids	Fish eye	Polyamid grid
Bend grids	Fish slot	Plastic grid
Blotted grids	Pop eye	PA-Rubber grid
Oval grids		
Hooped / Fixed angle		

The materials used for making hard BRDs include steel rods, aluminum rods, steel or aluminum tubing, fiberglass rods, polyamide grids etc. Over thirty different hard BRD designs have been developed for different resource groups and fishing areas.

Flat grid BRDs

Flat grid BRDs are mostly rectangular in shape without any bend in grid bars (Fig. 108). This type of design was developed in Norway originally to exclude jelly fish. The grid either made of aluminum or steel is usually mounted in the throat section at an angle of 45 to 50 degree from the horizontal. The grid is usually associated with an accelerator funnel for guiding the catch to the grid. Escape openings are provided either on top or bottom and are either kept open or covered with a flap of netting. Examples of flat grid BRDs are Nordmore grid, Wicks TED, Kelly / Girourard grid and short V-grid.

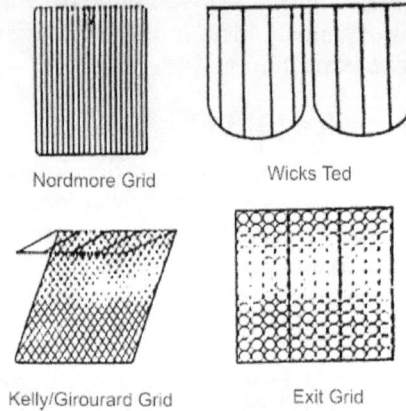

Nordmore Grid   Wicks Ted

Kelly/Girourard Grid   Exit Grid

Fig. 108: Flat grid BRDs

Bent grid BRDs

Bent grid BRDs are either rectangular or elliptical in shape. In this group of BRDs, welding cross bars or by leaving one end of the bars without joining to the frame. Steel aluminum and polyamide are used to construct the grids. The important grids under this category are Flounder TED, Jones TED, Matagorda, Hinged grid and Anthony Weedless. (Fig. 109).

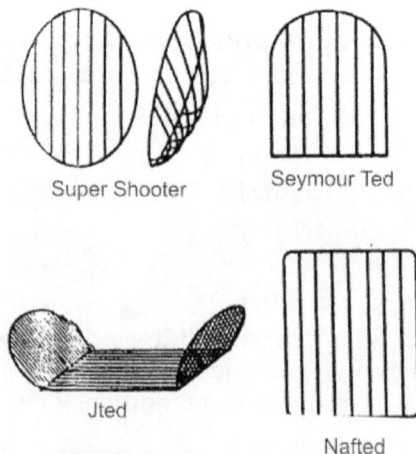

Super Shooter   Seymour Ted

Jted   Nafted

Fig. 109: Bent grid BRDs

Hooped and fixed angle grids

Hooped and fixed angle BRDs have circular, oval or rectangular hoops in front and rear of the deflecting grid, which is rigidly fixed in a frame work at the desired angle (Fig. 110). Materials used for construction are steel or aluminum. The main advantage of hoped TEDs are (a) sturdier construction for fishing in rugged conditions and (b) constant angle of deflector bars unaffected by changes in the elongation of netting. However, these designs are relatively cumbersome in terms of onboard handling and hence not in popular use. The NMFS Hooped BRD, Camerson shooter BRD and Fixed angle BRD comes under this category.

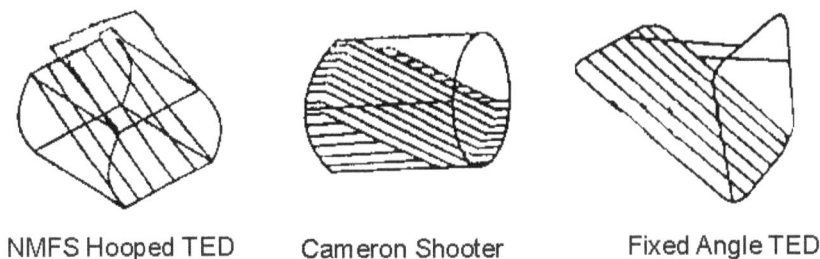

NMFS Hooped TED   Cameron Shooter   Fixed Angle TED

Fig. 110 : Hooped and Fixed angle BRDs

BRDs with rigid escape slots

BRDs with rigid escape slots are designed to facilitate the escapement of fish from the cod end (Fig. 111). Fish eye is the most important BRD coming under this category. It consists of an oval shaped rigid structure with 8-15 cm height and 30-40 cm width, with supporting frames made of stainless steel rods. Fishes swim backward from the cod end and escape through fish eye. There are several design variations of fish eye, such as, Florida fish eye (FFE) used in the Southeast U.S. Atlantic and in the Gulf of Mexico, Florida Fish Excluder (FPE) and Snake eye BRD used in North Carolina Bay. Fish eyes of different size and shape are used in South Atlantic and in the Gulf of Mexico. Fish slot, Sea Eagle BRD, Pop eye Fish Excluder or Fish box BRD, Ex-it and Sort-V grids are other designs in this category.

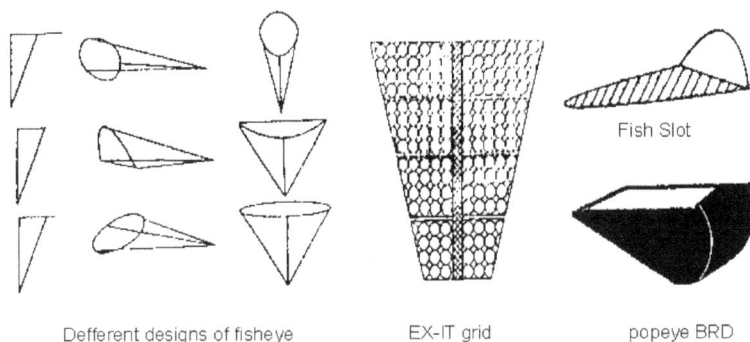

Defferent designs of fisheye   EX-IT grid   popeye BRD

Fish Slot

Fig. 111 : BRDs with rigid escape opening

Semi-flexible BRDs

Semi-flexible BRDs are constructed out of semi-flexible or flexible materials like, plastic, polyamide, FRP and rubber (Fig. 112). These include (a) flexible plastic grid made of polyethylene and the grid frame consisted of plastic tubes used in the North Sea brown shrimp fishery, (b) polyamide grid with hinges for operation from net drums used in the Danish experiments in the North Sea shrimp fishery and (c) Polyamide-rubber grid design from Denmark.

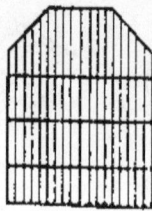

Polyamide Grid   Hinged polyamid Grdi

Plastic Guid   Pa-Rubber Grid

Fig. 112 : Semi-flexible BRDs

Combination BRDs: Some times, two or more BRDs are combined in a single gear to enhance the efficiency. Researchers proposed different combination of grids, slotted BRDs, such as, fish eye and soft BRDs, such as, square mesh window, big eye BRD and similar ones to obtain optimum results.

Thirty three hard BRD designs described have been operated either experimentally or commercially in different fishing areas with promising results. Super shooter TED operations indicated shrimp loss between 2 and 12 percent in Australian waters. The Super shooter TED also performed well in areas where the other inclined grid BRDs tended to clog due to accumulation of sponges and sea weeds and worked well when used in combination with other BRDs, such as, fish eye. (Fig. 113)

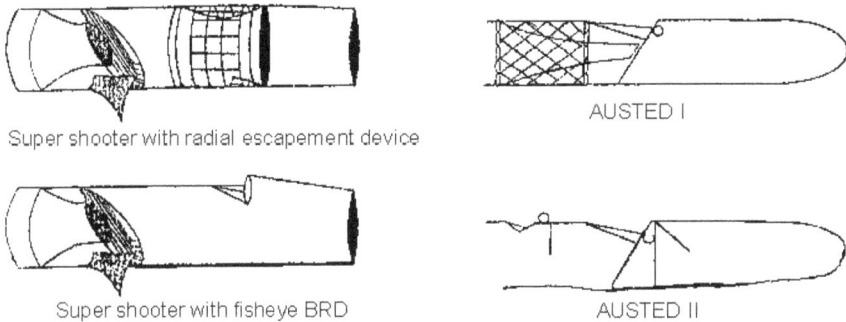

Super shooter with radial escapement device

AUSTED I

Super shooter with fisheye BRD

AUSTED II

Fig. 113 : Combination BRDs

As 40 percent of the by catch in India is contributed by juveniles, a Juvenile Fish Excluder-cum Shrimp Sorting Device (JFE-SSD) has been developed for bringing down the by catch of juveniles and small sized non-targeted species in commercial shrimp trawl. The JFE-SSD operations off Cochin, India have realized by catch reduction up to 43% with a shrimp retention 96-97%. Super shooter TED operations off Visakhapatnam, India indicating higher exclusion of fish when the exit on the upper side. In both cases, 100% escapement of turtles was observed. The NAFFED operations in Australian waters in combination with square mesh window during the commercial trials indicated shrimp loss of 3.3% in the catch of a standard trawl. CIFT-TED operations in Arabian Sea and Bay of Bengal indicated 100% exclusion of sea turtles with a mean catch loss in the range of 0.52 to 0.97% for shrimp and 2.44 to 3.27% for non-shrimp resources. The CIFT-TED was also reported as a simple BRD which can be fabricated easily and installed with minimum training, using net making skills and workshop facilities available locally.

Experiments with Nordmore grid in Norwegian waters, have shown a low and fairly constant shrimp loss of 2.5%, while fishes above 20 mm size were observd to escape. Experiments using Nordmore grid in Nova Scotia, Canada showed target catch loss of 2-5 percent and by catch reduction of 48-98 percent. Experiments with Nordmore grid in Portuguese continental waters showed up to 78.5 percent exclusion of large by catch species with negligible loss of target species catch. Experiments with modified version of Nordmore grids made of plastic in the North Sea reduced more than 70% fish and 65% benthos with a target catch loss of 15%. It was observed that escapement of juveniles up to 95% during experiments with two different rigid sorting grids, namely, Sort V-grid and Ex-it grid in coastal waters of Nambia.

Performance of fish eye depends on the shape, size, position, light and water current. Fish eye experiments conducted in Florida and coastal Australian waters showed enhanced by catch reduction when used in combination with other BRDs. Combination BRDs are used to increase the efficiency of the BRD in terms of by catch reduction and retention of target catch. Average by catch reduction ranged between 18 and 55 percent for AUSTED-I and between 15 and 49 percent for AUSTED-IT depending on fishing conditions in Australian waters, besides shrimp retention and exclusion of turtles and large animals.

Increased use of BRDs in trawling is an important reason contributing to the reduction in by catches in recent years .Cooperation between fishing industry, scientists and other stakeholders is fundamental for the success of by ccatch management efforts. Ease of construction and operation of the BRDs, cost effectiveness of the technology and the economic benefits influence the adoption of by catch reduction technologies. BRDs most appropriate to the regional fishing conditions should be adopted and enforced legally after careful scientific evaluation and commercial trials to ensure long term sustainability and to protect the biodiversity of fishery resources.

Soft by catch reduction devices

BRDs can be broadly classified into three categories based on the type of materials used for their construction, namely, Soft BRDs, Hard BRDs and combination BRDs. Hard BRDs are those which used hard or semi-flexible grids and structures for separation and excluding by catch. Combination BRDs use more than one BRD, usually hard BRD in combination with soft BRD, integrated to a single system.

The soft BRDs use soft structures made of netting and rope frames instead of rigid grids, prevalent in hard BRDs, for separating and excluding by catch. Based on the structure and principles of operation, they are classified into five categories, namely, (1) escape windows, (2) radial escapement section without funnel, (3) radial escapement section with funnel, (4) BRDs with differently shaped slits and (5) BRDs with guiding separator panel. Soft BRDs have advantages such as, ease of handling, low weight, simplicity in construction and low cost compared to hard BRDs.

Escape window function based on the differential behavior of fishes and shrimps. Fishes that have entered the cod end tend to swim back and escape when suitable excape windows are provided at the top in the front section of the cod end. Square mesh window (Fig. 114) and rope BRD (Fig. 115) are the examples of this category.

Fig. 114: Square mesh window

Fig. 115: Rope BRD

Using square mesh windows have indicated their effectiveness in reducing by catch by 30 to 40 percent in Northern prawn trawl fisheries, Australia. Square mesh has the advantage that the mesh opening is not distorted while under operation, unlike diamond meshes. Trial fishing conducted in Persian Gulf waters have shown that rope BRD is effective in excluding 25 percent of the by catch with no loss of shrimp or commercial fish species. Use of square mesh panels has been found to reduce the by catch particularly juveniles and young ones by about 20 percent in Indian waters.

In radial escapement section without funnel, a radial section of netting with large meshes is provided between hind belly and cod end. Small sized fishes, jelly fish and other by catch components, which have low swimming ability, are expelled due to enhanced water flow through large mesh section. Experiments in Japanese waters, using TREND were found to give safe escapement to juvenile fish with better opportunity for survival (Fig. 116).

Fig. 116: TREND

Radial escapement devices with funnel are positioned between hind belly and cod end of the trawl. A small meshed funnel accelerates the water flow inside the trawl and carries the catch towards the cod end. Actively swimming fishes swim back and escape through large mesh netting section surrounding the funnel, where the water flow rate is weak, while the shrimps are retained in the cod end (Fig. 117).

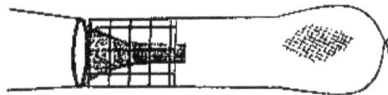

Fig. 117: Radial escapement device

Studies using radial escapement device have shown 20 to 40 percent reduction in the fish by catch in Australia's Northern prawn fishery. Studies in India have indicated 14 to 21 percent reduction in fish by catch by using designs of radial escapement device with 80 mm, 100 mm and 150 mm square meshes surrounding the funnel. Experiments in Louisiana have shown that nets fitted with extended funnel BRD (Fig. 118) and skirted extended funnel BRD caught lesser by catch than the control nets. The extended funnel BRD provided 44 percent fish reduction with 5 percent shrimp loss. The monofilament BRD used in commercial trawling has been reported to give 25 to 51 percent reduction in by catch, without problems of clogging. By catch reduction by Neil-Olsen BRD has been reported as 27 to 45 percent in tropical coastal waters.

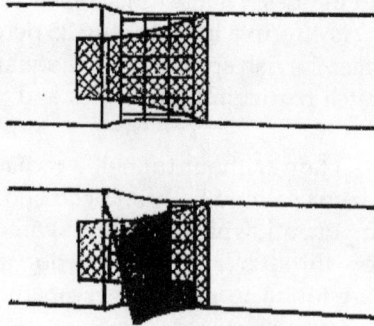

Fig. 118 : Extended funnel BRD

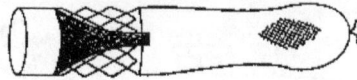

Fig. 119 : Monofilament BRD

BRDs with differently shaped slits, utilize the difference in behavior of fish and shrimp. Fishes that enter the cod end are given opportunity to swim back and escape by providing slits in the netting on the top side of the cod end, or hind belly, while shrimps are retained in the cod end. In diamond BRD, a diamond shaped hole is provided on the top of the cod end (Fig. 121).

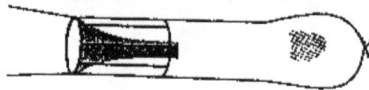

Fig. 120: Neil Olsen BRD

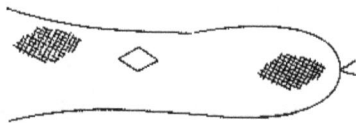

Fig. 121 : Diamond BRD

Average by catch reduction V-cut BRD operated in Queensland east coast trawl fishery has reported to be 16 percent with very low or no shrimp loss (Fig. 122) Investigations in India with big eye BRD have shown by catch exclusion in the range of 8 to 11 percent, while the same device reduces by catch by 30 to 40 percent intropical coastal waters in commercially operated shrimp fleet in Queensland east coast waters (Fig. 123).

Fig. 122 : V-cut BRD

Fig. 123 : Big eye BRD

Guiding or separator panels are used to achieve separation of the by catch by using differences in their behavior or size. BRDs with guiding panels lead the fishes to escape openings, making use of herding effect of the netting panels on fin fishes. The shrimps are not subjected to herding effect and hence passes through the meshes towards the cod end. BRDs with separating panels physically separate the catch according to the size, with the use of appropriate mesh sizes. Shrimps pass through the panels to the cod end, while by catch as, fishes and sea turtles are directed towards the exit opening. Seperator panel BRD operations in New South Wales shrimp trawl fisheries have indicated a shrimp loss of 2 to 30 percent and fish exclusion of 30 to 80 percent. Studies in India using separator panel BRD installed in shrimp trawl have shown a target catch loss of 44 to 53 percent due to vulnerability of the device to clogging leading to ineffectual sorting. Authement-Ledet BRD (Fig. 124) with bottom opening has been reported to give better exclusion of fishes, while top opening BRD entailed a minimum shrimp.

Fig. 124: Authement-Ledet Excluder

The Morrison TED (Fig. 125), Parker TED (Fig. 126) and Andrew TED (Fig. 127) are efficient soft TEDs which are used to exclude sea turtles and large marine animals in many countries.

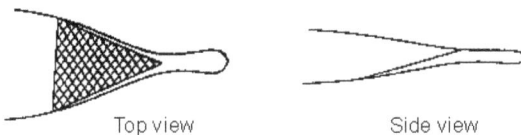

Top view   Side view

Fig. 125 : Morrison soft TED

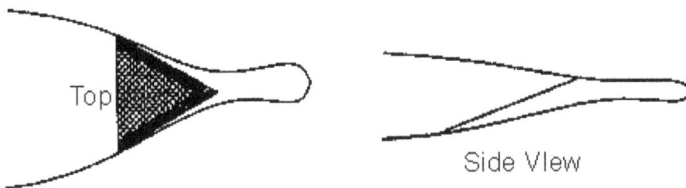

Top   Side View

Fig. 126 : Parker soft TED

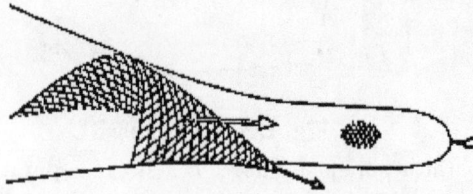

Fig. 127: Andrews soft TED

Proper installation of the soft TEDs are essential in order to ensure their efficient performance. Morrison soft TED has been used successfully to exclude sea turtles in Gulf of Mexico.

Investigations in Indian waters with sieve net (Fig. 128) have given 33 to 37 percent exclusion in by catch with minimum shrimp loss.

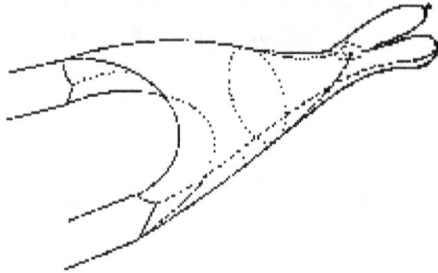

Fig. 128: Sieve net

By catch reduction has been taken as a serious issue in almost all the fishing nations. However, implementation of BRDs in different fishing areas has been desperate partly due to high catch loss and the difficulties in adapting the device to local conditions.Soft BRDs have been developed and tested in many countries, like, Australia, USA, Mexico, Belgium, Denmark, France and India. BRDs differ in their construction and performance based on the type of fishing and geographic peculiarities of the fishing grounds.

Soft BRDs have advantage, such as, ease of hauling, low weight and simplicity of construction and low cost compared to hard BRDs. Many of them such as, square mesh window, square mesh cod end, sieve net, radial escapement devices and their design variations and big eye BRD are popular among shrimp fishermen in different areas. An important draw back of soft BRDs is the vulnerability to clogging of the netting panels used in the construction due to gilling and tangling by fish or marine debris. The use of soft BRDs, such as, big eye, sieve net appropriate for Indian fishing conditions need to be promoted in order to support long-term sustainability of fishing resources and protection of biodiversity.

13

Power Requirement For Trawl Net Operation

There is a tendency among some of the fishing boat owners to over power their boats on the assumption that it will result in better performances of their boats. But the fact remains, in actual practice, trawlers over powered do not give any better performance than economically powered boat, on the other hand, over powering of the boat will weaken it. The performance of fishing boat is mainly dependent on its (a) speed; (b) load carrying capacity; (c) stability and (d) ease of fishing operation.

As far as the speed of the fishing boat is concerned, it is governed by the following relation;

$$V = 1.4 \sqrt{L}$$

where,

"V" is the speed of the boat in knots and "L" is the water line length of the boat in feet. For example, a boat with water line length of 36 feet can develop a maximum speed of about 8.4 knots, provided all other factors, such as, the angle of entrance, prismatic coefficient etc are properly chosen so as to give this speed. Otherwise the speed will be less than this. A study of the various fishing boats operating all over the world will amply prove the validity of the above relation.

For example, two Dutch fishing vessel of 54 feet LOA were constructed using one and the same drawing. The first vessel had a fish hold capacity of 1000 cubic feet and was fitted with 60 HP engine. It developed a speed of 8 knots. The second vessel built after this was fitted with 80 HP engine and it developed the same speed of 8 knots though the fish hold capacity in this case was reduced to 750 cubic feet. In this case,

$$V / \sqrt{L} = 1.1 \text{ was used}$$

Thus greater speeds than given by the above equation can not be obtained with the displacement hulls of the fishing boats. If more speed is to be obtained, the entire design of the hull has to be modified to create semi-planning or planning action. The crafts for sking or racing, where speed is the only criterion, are built with planning hulls. Such planning hulls can develop substantially higher speeds than displacement hulls, but their load carrying capacity is very much less and hence they are not suitable for fishing boats. Thus fishing boats, if powered with engines capable of delivering enough power as to develop speed determined by the above formula will be working economically. Any attempt to deliver more power than this will make the stern to sunk and the bow to rise and this tendency will increase as more and more power is delivered. That is, the boat will try to behave like a semi-planning hull, which it is not. Hence the boat will not be floating on its designed water line, whereas it is necessary for any boat to float on its designed water line or of water line parallel to it for better overall performance. Further such a trimming of the boat aft will affect the boat in the following ways.

1. The wooden boat is normally seathed up to load water line. When the boat is sunk above the water line at aft portion, the unseathed portion of the hull will also come in contact with the water, giving way for the deterioration of the wood by means of borers and insects.

2. In small boats, the engine is inclined up to a maximum of 12 degree for obtaining submergence of the propeller. As the boat sinks further down the stern side, the inclination of the engine to the horizontal plane increases. As a result of this, the engine may develop one or more of the following troubles.

   (i) The lubrication system may not work well, as the lubricant is collected on one side of the sump, thus starving the pump. Any shortage in the supply of lubricant to the moving parts of the engine will immediately affect them and at times may even lead to major break downs.

   (ii) Since the capacity of the lubrication pump is also limited, it is likely that some of the parts at the far end of the pump may not receive the lubricant at all, with the result that the wear and tear on these components will be more.

   (iii) The engine may start vibrating severely as the forces acting on the engine are subjected to a change of direction.

In short, the performance of the engine will be seriously affected. Thus it will be clear that over powering does not make the boat move faster. On the other hand, it harms the engine and the boat.

The dead weight or the load carrying capacity of a boat is also decided during the design stage itself. Further in a fishing boat, the ice and and fish are carried in the fish hold whose capacity is already limited. Any attempt to overload the boat will make the boat sink below the load water line. This will result in a reduction in the free board of the vessel. This is very harmful for the safety of the boat.

Stability is a "sine qua non" for any floating vessel. In fact, inadequate stability could be ascribed as the reason for many a failure of boats in the seas. Over powering as will be seen, will rob the boat of its seaworthiness and efficiency. The seaworthiness of a boat is decided by the ability of the boat to stand rough weather etc. To meet these, a boat should be neither "tender" nor "stiff". A tender vessel will be easily subjected to rolling and will therefore require a constant operation of the helm. On the other hand, a stiff vessel will have little "sea kindliness" and will give maneuvering difficulties. A compromise must therefore be arrived at in deciding the stability of a vessel. The only tool by which this can be measured is the value of GM (the distance between the center of gravity and meta-center). It has been found from experiences with fishing boats that a value of two feet GM results in satisfactory stability conditions. Hence as long as the boat floats in the designed water line, or any line parallel to it, the resulting value of GM can be considered to be safe. But any boat over powered will not float on the designed water line but sinks at aft side. Thus it is quite likely that the value of GM may fall below the safe limit, with a consequence danger to the vessel. Hence there is no doubt that over powering makes the boat unstable and unsafe.

A fishing boat is essentially a platform over which all the fishing activities are carried out. Any difficulty to perform this primary function indicates the poor quality of the boat. A boat should have (a) seaworthiness and (b) sea kindliness to carry out this. The boat that is over powered does not possess mese two qualities sufficiently. Further, a boat which is sunk at the stem side is likely to sink further down during trawling (stern trawling), which will not only result in discomfort, but also cause a host of other difficulties to the crew.

Over powering thus invite danger to the boat. Above all factors, "seaworthiness" must be given priority, because any attempt to undermine, it will endanger the safety of the crew and the boat. Over powering has a devastating effect on boat's stability. Beyond certain limit, which can be called as the "economic limit" the engine becomes a destroyer instead of a producer.

Power requirement and supply to operate auxiliaries in fishing trawlers.

In big fishing trawlers, electric power supply from alternator is available to operate various auxiliaries located any where in the vessel. But the small crafts are entirely dependent on one source of power, namely, the main diesel engine- To utilize this source of motive power to operate winches and other ausilliaries, mechanical drive involving shafts, bearings and pulleys is inconvenient. The use of hydraulics for such operations seems to be an ideal solution. For instance, one pump may be utilized for operating trawl and another for winches in parallel. Additional operation of steering system, power blocks etc. can be achieved by adding few components to the basic system,

Advantage of hydraulic system over mechanical drive.

(a) Hydraulic system is enclosed and more compact compared to mechanical drive from the engine. Actuator (hydraulic motor or cylinder) and control valve on the deck can be completely enclosed for weather protection.

(b) No rigid drives are involved and there is no stress on mechanical component due to vibration and buckling of the craft. Hydraulic piping can be laid to suit space and contour of the structure to cause minimum interference.

(c) The hydraulic requires the lowest possible reaction time with high starting torque and infinite control of speed, irrespective of engine speed.

(d) Very little or no maintenance is required for a properly installed system.

(e) System is self protected against over load and would stall against over load with out causing failure of mechanical component.

(f) All components are self lubricated by the fluid medium, hence longer life.

(g) Engagement of clutch is easier because of inching capability in either direction of rotation.

(h) Operation is silent, which is a great advantage on modern vessels.

Hydraulic system for winch

The basic hydraulic system consists of (a) Hydraulic oil reservoir, (b) Engine drive pump; (c) Relief valve; (d) Control valve; (e) Counter balance or brake valve; (f) Check valve; (g) Hydraulic motor and (h) Inter connecting pipes and hoses.

A system where the winch is designed for uni-directional operation (for handling only), brake valve and check valve are not required.

Two types of hydraulic system are in use at present; (i) Low pressure and high volume system and (ii) High pressure and low volume system. Each of the two types of systems has its own advantage and disadvantage. Low pressure and high volume systems are preferred by certain European and Japanese concerns, whereas high pressure and low volume systems are in vogue in USA and Latin American countries — High pressure and low volume systems are about 25% cheaper than high volume low pressure systems.

Comparison between high pressure low volume and low pressure high volume system

High pressure low volume system	Low pressure high volume system
1. Cheaper	Costlier
2. Lighter	Heavier
3. Needs reduction gear at winch	Needs reduction gear at pump
4. More accuracy of components required	Does not require much accuracy
5. Smaller piping	Bigger piping
6. Compact	Bulky
7. Easier servicing, cartridge construction	On board servicing is difficult
8. Shorter life of fluid medium	Longer life of fluid medium
9. Leakage possibilities more	Leakage possibilities less

Considering the above points of advantages and disadvantages, the high pressure and low volume systems are preferable for small fishing vessels due to restricted space availability.

With the present range of availability of hydraulic units available in the country, it is possible to design winches up to 2 tonnes pull with necessary hauling speed suitable for crafts up to 12.19 m length. Vessels up to 17.5 m can also be equipped with hydraulic winches from indigenous source except for the hydraulic motor, which has to be imported. For vessels beyond 17.5 m length, both hydraulic pumps and motors are to be imported. However, the rest of the mechanical components of the winch are well within the capacity of present winch manufacturers. This way import of complete winch can be avoided, thereby saving at least 85% of the winch cost in foreign exchange.

14

Electrical Trawl

Electric fishing was established early in 20th century and has continued to grow popularity. It is used successfully in various habitats and environments and is an accepted method for both commercial fishing and fisheries research world wide.

The principal fishery purpose for which electricity is used are undoubtedly those that involve fish capture. Electric fishing can be particularly useful in situations, where other techniques may be ineffective owing to the nature of either the target species or the habitat. Some fish, such as, eel, tench are difficult to catch and some habitats, such as, weed beds, rocky shores arid fast flowing water are difficult to fish. Electrical gear can either catch fish directly or influence their capture by enhancing other methods.

It is possible to catch fish with any kind of electric current, but to improve the efficiency, to avoid injury to the fish or to fish in the difficult condition ,the type of current is important. Alternating current (AC) tetanizes the fish and kills a high percentage of the catch, whereas direct current (DC) attracts the fish to the anode *(anodic galvanotaxis)* making capture easy and is least damaging. The effects of interrupted or pulsed direct current (pulsed DC) are intermediate between AC and DC.

The combinations of pulsed current are practically infinite, as they are characterized by three parameters which have a high degree of flexibility in pulse shape, pulse duration and pulse frequency. The problem arises as to how to evaluate the optimum of each parameter to ensure the most successful and efficient fishing mode. Considerable work has been oriented towards this objective by varying one parameter and keeping others constant. Unfortunately, this has led to differing conclusions. For example, on some occasions a low frequency appears best, whilst, on others, a high frequency is beneficial, or condenser discharges produce negligible injury in certain circumstances, but not in others.

In sea fisheries, the use of electrical gear is developed. It is possible to concentrate fish shoal by keeping an electrode before the aperture of a trawl net and to paralyze them so that they can not escape the trawl. An electrode placed in front of a pelagic trawl (mid-water trawl) will attract fish shoal even from the lower area of the electrical field and guide them into the catching area.(Fig. 129). This means, fish from the region beneath the net, which would otherwise not be taken, will be caught. An electrical mid-water trawl also enables the otherwise "unfishable" uneven bottom to be fished as the fish living near the ground are attracted towards the electrode.

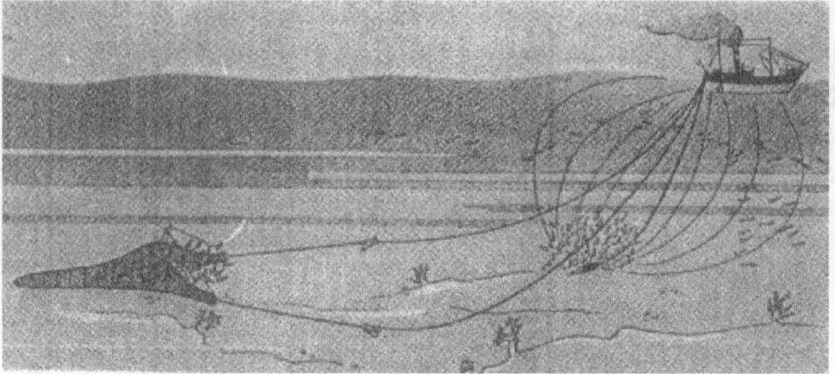

Fig. 129: Schematic diagram of an electric trawl.

The development of useful electrical fishing gear in sea encountered many difficulties. First, the greater conductivity of sea water, as compared with fresh water had to be overcome by inboard equipment and second, the electrical fields and ranges had to be considerably larger. Electrical fishing gear for deep sea fisheries, is of course, only of interest when it promises substantially increased catches. Such gear may not require much space aboard nor complicate usual fishing methods.

Impulse generator for electrical trawl

The electro-technical solution of the problem was achieved by Kreutzer in cooperation with the *Bundesforschungsanstalt fur Fishcherei,* Hamburg and the Siemens-Schuckert=Werke AG by developing an impulse generator (the ignitron valves were supplied by AEG, Hamburg).

A condenser, which is loaded relatively slowly by a direct current generator is discharged within a very short time over a throttle and passes the concentrated energy over the electrodes into the water to obtain the high peak voltages required. Adjustment of loading and discharging is done by means of an electronic governor, whereby the discharge current is switched over two parallely connected ignitron. The principle of the production of the pulsating current is described in Fig. 130.

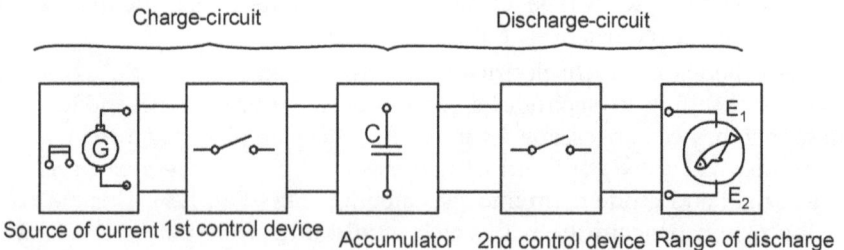

Fig. 130 : Basic principle of producing impulse current (according to Haier)

The detailed, technical designs and working method of impulse generator is given in Haier's book, "*Die elektrotechnischen Grundlagen der Elektrofischerei der Meer*" (The Electro-Technical bases of Electrical Fishing in the Sea).

Experiments made in 1952 in the North Sea and the Bay of Kiel with the minesweeper R 96 revealed that it is possible to produce current impulses of at least 10000 amperes. The average capacity amounted to about 180 to 200 kilowatts.

To harness the energy employed for the production of pulsating currents in a practical way, it is necessary to achieve the required density of pulsating current within as large a range as possible, without wasting too much energy in the direct vicinity of the electrodes. According to Haier, that can be reached by various measures;

1. By forming electrodes with largest possible surfaces;
2. By concentrating the potential at individual points or areas of the field and by giving the current a certain direction by means of auxiliary electrodes;
3. By cumulating the current through short impulses for orienting the field.

The rate of impulses can be varied, according to the species offish, from 3 to 60 impulses per second. The length of the impulses can be changed in the ratio of about 1:3. Subsequently, the generator used on board R 96 was replaced by new impulse generator, designed by Kreutzer.

The application of the generator is more useful in sea fisheries. Such a generator for instance, can be used in the trawl fishery. Fish shoals can be concentrated before the net opening by means of anodic effect, and paralysis prevents the fish escaping the approaching net providing the ratio between the current flow and interruption is correlated to the towing speed. In this way it should be possible to achieve considerable increases in catches.

The generator installed in R 96 was tested as to its anodic effect on sea fish. The gear required for electrical trawl fishing was designed and tested, together with the electrode designed for this experiments (Fig. 131).The catching electrode (2) was installed in front of the mouth of the trawl net (1) and the electrical energy transmitted to the electrode by a cable (3).

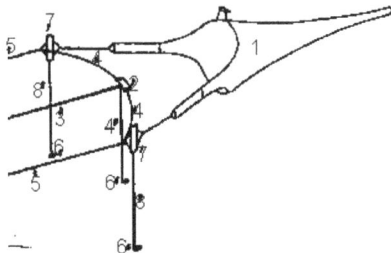

Fig. 131 : Diagram of an electrical trawl net.

An electrode with greatest possible surface and a stabilizing device is put before the opening of a mid-water trawl. A conducting cable connects the electrode to the electrical gear on the vessel. A line on the way lines ensures an additional adjustment of the electrode. The line serves simultaneously for lowering the electrode. This is done when the trawl has taken its fishing position. Three iron weights and lines fastened to them and to the trawl boards ensure the unchanged distance of the fishing gear from the bottom of the sea.

Shape of electrode for sea fishing trawl.

The electrode designed by Suberkrub, after thorough experiments has a total surface of 5 square meters (54 sq. feet) (Fig. 132).

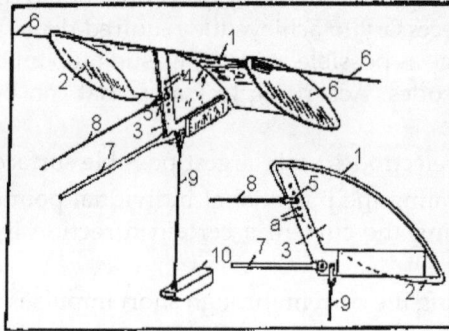

Fig. 132 : Catching electrode of an electrical trawl (according to Suberkrub)

The use of electrodes with large surface is of great importance with regard to the required energy, as can be understood from Haier s investigations. A function of shape and surface area of the electrode is the direct current spreading resistance. This decreases approximately corresponding to the square root of the surface. For a specific resistance of 25 Ohms x centimeter, it remains 20 Ohms x centimeter for a diameter of 2 meters (6 and half feet) so that the loss of potential near the electrodes is great. Figure 133 shows the measured field between a plate electrode with a surface of 4 square meters (43 sq, feet) and an electrode of 20 square meters (215 sq. feet) size attached to the vessel. According to this diagram the potential drop is 60 percent at a distance of 2 meters (6 and half feet), whereas 40 percent of the voltage would suffice if the size of the electrode were tripled.

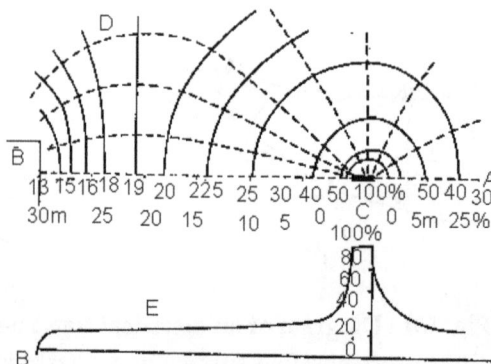

Fig. 133 : Field of an experimental arrangement (according to Haier)

The opinions on the suitable design of an electrode may vary.The closed spherical shape of the surface of the electrode, mentioned above, is usually impossible because of unwieldiness and the great resistance experienced in towing. A float was used which, owing to its hydrodynamic properties, had the smallest possible weight, the smallest possible resistance to the water and which took the position opposite the net opening without any difficulties, Perforated bodies can be chosen instead of a closed electrical surface. These need not be rigid. They develop their size in the water by the influence of weight, buoyancy or resistance (Fig. 134).

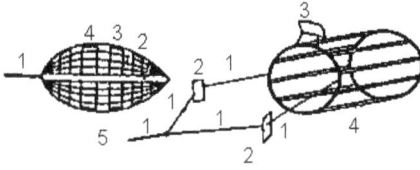

Fig. 134 : Shape of electrodes with perforated surface (according to Haier)

Haier thinks that it is possible to improve the development and consequently, the effect of electrical field by using auxiliary electrodes. These auxiliary electrodes should govern the potential and provide the area around them with a greater intermediary potential (Fig. 135).

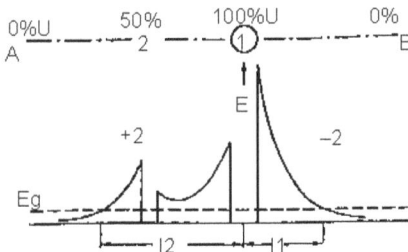

Fig. 135 : Control of the potential passing through the auxiliary electrodes.

This intermediary potential can be produced by either linking the auxiliary electrodes to special potentiometer connections of the generator or by increasing the most effective conducting or spreading resistance beyond the resistance of the main electrode. In figure 136, metal trawl boards act as auxiliary electrodes.

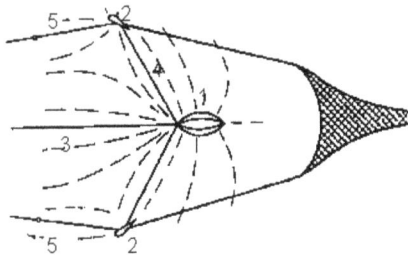

Fig. 136 : Arrangements of electrodes for a trawl net (according to Haier)

As is known, in the alternating current technique, quick changeable currents in a medium, through which the conducting cable runs, tend to flow back along the cable. It can be supposed in this case, that the current in Fig. 136 would also flow back from the electrodes along the conducting cable (3). Such an effect would considerably increase the efficiency of the pulse generator in fisheries as suggested by Kreutzer.

In order to keep the potential drop near the catching electrode, as small as possible, the electrodes must be very large. The Ohm resistance of the whole arrangement will thus be reduced. Since only the strength of the electrical current in the water is decisive for the range of fishing effect, the potential can be reduced in case of a small resistance and consequently, also the required capacity. But even here a limit will soon be reached. For a quadruple increase of the surface area of the electrodes results in only a 50 percent reduction of the spreading resistance of the electrode. One must attempt to restrict the unlimited expansion of the current in the marginal zones of the fishing area as well as to avoid the great potential drop and thus high consumption of energy near the catching electrode. The hull of the vessel as cathode is usually large enough not to cause any undesirable losses.

These various advantages are obtained by a special arrangement of the electrodes and conducting cables. The invention is based on the following observations. If an electrode fastened to a long cable is lowered from a vessel and the cable arranged in the form of a circle (as shown in A, Fig. 137) the current expands, if a direct current is used, corresponding to the inserted field lines. In this case a wooden vessel with a large electrode beneath the stem was used. The resistance of the whole arrangement is equal to the sum of the resistance of cable and the spreading resistance of the two electrodes. Quite different are the condition with alternating current or with short rated direct current pulses of equal or changing direction. Owing to the large extension of the arrangement, the self induction formed by the cable loop is so great that the inductive resistance becomes large as compared with the resistance of the arrangement in case of direct current. Unlike direct current, the current does not in this case take the shortest way to the other electrode. The higher the frequency of the employed alternating current, or the shorter direct current pulses used are, the nearer the generated current lines will approach the cable. The Ohm resistance of the water column along the cable passed by the current is, in consequence of the great electrical inductive resistance of the sea water, still small as compared with the inductive resistance of the whole arrangement. The current passing through the cable and that returning along the cable behave like a bifilar winding, the inductiveness of which is approximately zero. The resistance of the scheme is then determined by the sum of the spreading resistance of the two electrodes, the resistance of the cable and the Ohm resistance of the water column. The wide expansion of the current is thereby prevented. The current density along the whole way between the electrodes can be made so great that the whole circuit has a fishing effect.

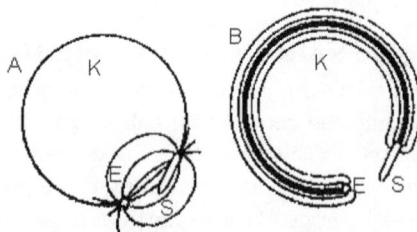

Fig. 137: Catching electrodes arranged in loops (according to Kreutzer). Extension of the electric current in sea water.

This increases substantially the electrically influenced area. The occurring field is homogeneous in the direction of the cable, and at right angles to the cable, the potential of the field is continuously increasing. When a double loop is formed, like a coil with two spirals, then the current also returns in two loops so that the bifilar character of the scheme is maintained. The current density in the water is then double that issuing from one electrode. By means of a cable coil with two spirals, the losses near the electrodes are reduced to a quarter as compared with only one loop of the cable for the same voltage in the homogeneous part of the field.

When using two or more loops, it is possible to reduce the space, but yet, to meet the requirement, that the inductive resistance of the coil is great enough as compared with the direct current resistance of the arrangement.

The limit for the height of current pulses is about 20 kilo amperes. Higher current peaks are not possible, as the normally used ignitrons do not permit higher peaks. If a current peak of the double amount should be attained, the potential would have to be doubled using the same arrangement.

When using a double loop of cable, the potential must be only slightly increased to produce a double current density in the water. The greatest Ohm resistance-at the catching electrode- is in this case only passed by half the amount of current flowing through the water. If it is desired to produce a blocking effect towards the depth, the second cable loop can be put into greater depths than the first one.

If this arrangement of electrodes is satisfactory, remarkable results can be obtained.

The energy required for this gear can be produced onboard fishing vessels, such as, trawlers.

15 Instrumentation of Trawl Gear

Instrument in trawl gear performance

In the case of a trawl net under operation, there are many hydrodynamic forces acting on it. Since the trawl net is operated much below the water surface, visual observations is rather difficult. Model testing is an alternative method. But as a model can not be an exact replica of the original in all respects, model testing has only a limited application and an ultimate test with the actual gear in the true field conditions is necessary.

Instrumentation is the best resort for direct measurement of hydrodynamic parameters, like, water currents, uneven grounds fairly accurate and easily. The amazing progress of physical sciences is attributed to the wide possibility of instrumentation in these fields. The various electronic techniques are widely used for the conversion of the signals and their easy transmission to the required quarter.

Underwater measurement

Basically there are three methods for obtaining measurements underwater .They are (i) to record the signals in underwater units and read them after hauling in (II) to convert these information into electrical signals and transmit them to the vessel by means of electrical cables and (iii) to convert them into ultra-sonic signals and transmit to the vessel, where they will be received and translated into corresponding values. In the first method the necessary instruments will be heavy and bulky. The use of heavy instruments in the fishing gear can disturb its natural functions and is also risky that such costly equipments may be damaged in the rough conditions in the sea. Also the measurements can be ascertained only after hauling the gear and the time of operation is also limited to one rotation of the drum. The second method, does not in any way affect the natural performance of the gear, since, the necessary transducers required for converting the parameters into electrical signals can be made light and no recorders or such other heavy parts are required to be fixed on the gear. The third method possesses the combined merits of the first and second, i.e.no wire is required for transmitting the signals and measurements are instantaneous and continuous. But such ultra-sonic transmitters are also heavy, as they have to include the necessary electronic components and the heavy nickel or crystal transducer for converting the signals into ultra-sonic waves, in addition to the batteries necessary for driving the whole mechanism.

Measurement of working depth of the trawl gear

Hamuro and Ishii (1964 a) in Japan developed a heavy and large recording type instrument, which records the depth of operation of the gear on a rotating paper drum. A hydrostatic pressure cell picks up the pressure and a pointer moves over a rotating drum. The measurement is obtained after the gear is hauled up. The whole system is attached to the otter board. Hamuro and Ishii (1964 b) also described an ultrasonic transmitter for the same measurement. In this transmitter the hydrostatic pressure cells pick up the pressure and a contactor is made to move over a rotating drum, where the stylus movement is made proportional to the length of electrical pulses. These pulses are further converted into ultrasonic waves of various pulse lengths. The signals are received and displayed either in a recorder or in a micro-ammeter. Mc Neely (1959) described a potentiometer type pressure sensing transducer. A potentiometer is actuated by the movement of a diaphragm in the presence of hydrostatic pressure which is proportional to the depth and the variations in resistance of the potentiometer are conveyed to the boat by means of three core electric cable. A wireless telemetering type depth indicator developed by National Institute of Oceanography Survey, is reported in World Fishing (1963). In this instrument the frequency of an ultrasonic oscillator is controlled by the hydrostatic pressure. These frequency-modulated ultrasonic waves are transmitted to the boat, where they are received through a hydrophone. The depth is read in a calibrated dial by tuning the instrument to the received frequency. When it is off-tuned, a whistle is heard and the tuning position is indicated by the minimum pitch. Sivadas (1968 a) developed a mercury filled transducer for the above measurement. The transducer converts the hydrostatic pressure into variations in electrical resistance. These signals are conveyed to the boat by means of a two core cable. Sivadas (1969 c) has developed yet another electronic type instrument using solid state components. This instrument has got the additional advantage that the range of operation can be altered conveniently.

In all the above instruments, the basis of measurement is the hydrostatic pressure, which is proportional to the depth of the water. In the absence of instruments, specially developed for the purpose, some alternative methods were described by Mohr (1964) and Sharfe (1864) using echo-sounders. The transducer of the echo-sounder is tied on the head rope of the trawl net and kept facing vertically downwards. Another echo-sounder was operated from the boat measuring the depth of the sea bottom. Both of them are recorded separately and the difference between these gives the depth of operation of the gear. But this method has got the following defects;

1. If the sea bottom is uneven or is not parallel to the water surface, the difference in the sea bottom depths corresponding to the position of fishing gear and that of the boat will introduce error in the measurement. The magnitude of the error is very often significant.

2. The transducer in the gear requires special hydrofoil or suitable arrangements for keeping it always directed exactly down wards .This makes the transducer and the assembly quite heavy and bulky.

3. The long insulated shielded and seathed thick cable of the transducer makes the operation difficult.

Scharfe (1959 a) has measured this depth using two boats, the rear boat carrying an echo-sounder vertically above the trawl net which is towed by the other. The uncertainty of the exact position of the net produces errors in this measurement

Fig. 138: Electronic operational depth measuring instrument.

Warp tension

Another important variable in the trawl net is the resistance offered by the gear system, which can be measured in terms of tension on the warps. This tension is a combination of resistances offered by the warp, the otter boards and the net which includes floats, sinkers etc. It has often been noted that the warp tension is unsteady due to many factors, such as, pitching and rolling of the boat, sliding of the otter board and the net over the muddy and rough ground, unsteady movement of the otter boards etc. Measurement of warp tension helps one to judge the comparative efficiency of the nets, so far as resistance to motion through is concerned. A timely indication of the overloading or entanglement of the gear at the bottom helps the skipper to control the situation and save the costly net. Many of the abnormal behaviors of the gear will be indicated in the instrument.

Basically there are two types warp load meters. In one type, the full tension is applied along a hydraulic load cell or spring, while in the second method, only a portion of the actual load is given to the cell. The instruments developed hitherto have used one of the following three properties, namely, hydrostatic pressure developed in a cell due to tension, mechanical compression of a helical spring and electrical resistance variations in a strain gauge due to mechanical strain.

Scharfe (1959) used dial spring balances kept on board the vessel and measured the tension directly on the dial by connecting one end of the dial spring balance to the boat and the other end to the warp. Hamuro and Ishii (1964 a) applied the full tension on a spring and converted the compression of the spring into electrical variations and conveyed the signals to a convenient position when it is read in a micro-manometer. Carrothers (1968) described a full load hydrostatic type of warp tension recorder consisting of hydraulic cylinder pressure cell and a strip chart pressure recorder.

Unlike the full load tension meters, the partially loaded types possess the advantage that they can be easily fitted on the warp and used even while trawling. But they suffer from the main drawback that the response from the respective transducers are not linear to the actual tensions, reducing the accuracy towards higher values of tensions. This is because the tension which is effectively applied across the cell is 2T Sin A, where T is the actual tension and A is the deflection angle.

Satvanaravana and Nair (1965) have developed a deflection type, partially loaded portable tension meter, where the compression of a helical spring due to the partial tension across it is graduated in terms of full tension. Nichotts (1964) has converted the partial tension into electrical variations using strain gauges in a deflection type tension meter. The electrical variations conveyed to the wheel house were recorded in a continuous recorder. A partially loaded type electrical tension meter fitted in the winch developed by "France Peche" is reported by Foster (1958).

The tension is converted to electrical variations which are conveyed to the wheel house and displayed in an electronic continuous recorder. White Fish Authority developed a deflection type "ship installed' tension meter as reported by Driver and Ellis (1968). Here also the signals corresponding to tension are transmitted to the wheel house and recorded. The mechanical deflection type tension meter described in "World Fishing" is worth mentioning in this context. The small bending produced in a bar of mechanical deflection type portable instrument is mechanically amplified and displayed in a dial type micro-meter. The deflection angle is practically constant in this instrument, since the strain produced in the system is small and does not affect the angle significantly.

In the deflection type tension meters using strain gauges, the tension displayed in meters will be linear, because the deflection angle is practically constant. A wide variation in the deflection angle makes the response curve more non-linear producing less sensitivity towards smaller values of the angle. The ship installed deflection type warp load meters have overcome many of the disadvantages of the other types. Here the load cell is connected between the winch and the pully in the gallows, so that it possesses a large deflection base and the variation in deflection angle is neglected. This can be a permanent fitting in the deck, A major error in this instrument is due to the unsteady deflection angle owing to the varying diameter of the winch drum.

Tension measurement (under water)

There are different contradicting opinions (de Boer, 1959) among the research workers and technicians regarding the total resistance to motion of the gear system in relation to the catch. It has not yet been established as to whether the total resistance is increased or decreased. An increase in the resistance of the trawl net with catch is normally expected. But often times , the increase in tension due to catch is negligible compared to the decrease in tension due to the other factors, described below, contributing to it. As the catch increases, the tension contributed by the cod end increases considerably with the result an extra pull is induced through the length of the net. This affects the geometry of the net and for establishing the equilibrium, the otter boards will have to come closer. Further, as the otter boards come closer, the angle of attack is also reduced. Consequently, the resistance to motion of the whole gear system is reduced. An exact and authoritative knowledge of the above performance can be obtained only by measuring the various parameters such as, (1) the total warp tension, (2) the tension proportional to the catch, (3) the tension in between the otter boards and the net, (4) the horizontal opening and (5) the angle of attack. Underwater tension meters are intended for the measurement of the resistance to motion of various parts of a trawl net.

Basically the underwater line tension meters are of the same principle as that of warp tension meters. These instruments are to be operated underwater and the information should be either stored in water tight and light cases or communicated to the boat either by electric cable or by ultrasonic waves.

The underwater line tension meters are all based on hydrostatic principles, while the one developed Carrothers (1966) is of electronic type using semiconductor strain gauges. Since all the above types are underwater recorders, naturally they are heavy and bulky and often times the use of such heavy instruments is undesirable since the natural performance of the gear system is affected by them. Further, the required information can be obtained only after hauling the gear. The electronic type underwater tension meter developed by Sivadas (1970) using solid state component is comparatively light and rugged and it communicates the information to the boat instantaneously.

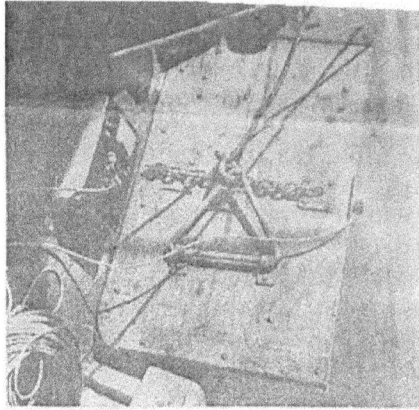

Fig. 139: Underwater tension telemeter

Measurement of angle of attack

Angle of attack is the angle between the plane of otter boards and their direction of motion. The resistance offered by the otter board and its efficiency are related to the angle of attack. Theoretically for the maximum efficiency the angle of attack has been estimated to be between 30 and 40 degree for rectangular type otter boards. There are many limitations in the calculations of the required angle, especially in the case of special otter boards, such as, curved, "L" shaped etc. The whole hydrodynamic system will maintain equilibrium if the parameters are within a limited range. If they exceeds the limits, the whole system is upset and does not function in the normal way. The measurement of the angle of attack is necessary to ascertain whether the otter board is operating under the required optimum range and also to estimate the most suitable range of this parameter, wherever other methods, such as, theoretical calculations are impossible.

Basically there are two types of angle of attack meters in use. The one developed by de Boer is mechanical under water recording type using curved board pasted with paper as the recording medium. The rigid recording part is fixed in the rear of the otter board and an area capable of free movement in any plane is allowed to slide along the bottom. At the opposite end of the arm, a pen is attached, the point of which is in contact with the paper board. The paper board is semi-circular with its center of curvature coinciding with the fulcrum of the arm. The clock-wise-driven pen moves in the vertical direction. As the otter board moves forward, the arm keeps itself as a reference direction to that of the movement of the otter board. Here the major disadvantages are that (a) the reference rod is to be always in contact with the bottom, which indirectly means that its use is restricted to bottom trawling, (b) being an underwater recording type instrument, the time of operation is limited and the information can be obtained only after hauling the gear. Further, as a heavy instrument, its use is limited to large size trawlers only.

The second type belongs to the one developed by Sivadas (1969 a). A small circular coil fixed on a brass plate is mounted on the top edge of the otter board. A plastic fin capable of free movement and fixed with a mild steel curved core is provided with its point of rotation coinciding with the center of curvature of the arc of the coil. As the otter board moves forward, the fin aligns itself in the direction of motion of the otter board, thereby establishing a reference to the direction of motion of otter board. Any change in the angle of otter board is indicated in terms of the angle of attack with the help of the coil and the mild steel core, the relative movement of which being proportional to the length of the latter. The signals are communicated to boat where they are translated into angle of attack. This instrument is very light (400 g) and gives an instantaneous measurement quite accurately.

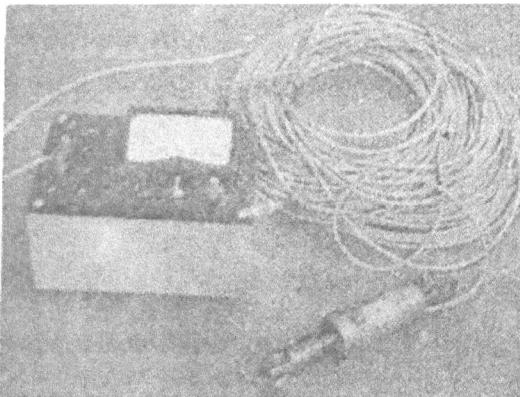

Fig. 140: Angle of attack meter Tilt measurements

Tilt Measurements

There are two possible tilts for an otter board. The tilt is the angle of the otter board around a horizontal axis parallel to its plane is usually called tilt. The tilt in angle of otter board around a horizontal axis perpendicular to its plane is usually called fore and aft tilt of the otter board. The tilt of the otter board is to be maintained within a limited range for the equilibrium of the gear system as well as for obtaining best results. The optimum value can be obtained by correlating the same with the other factors of the gear system, such as, horizontal opening, vertical opening, trawl resistance etc. Similarly the fore and aft tilt also is important. Fore and aft tilt in the negative direction is very dangerous, as it causes the otter boards to plough into the mud resulting in the breaking of the warps and loss of the gear.

The mechanical underwater self recording "clinometer" developed by de Boer (1959 a) is too heavy and bulky for a medium sized trawl net. A suspended heavy weight keeps its position always vertical and a pen attached on it records the tilt on a moving paper board. The same clinometer is used for both the tilt and fore and aft tilt by mounting them in two different ways.

The tilt meter developed by Sivadas (1969 b) is a telemetering type instrument with its very compact, light and rugged transducer. Another direct reading instrument based on the same principle has been made by Sivadas (1968 b) for the fore and aft tilt of the otter boards. Eventhough they work on the same principle, they differ in design and construction to suit the particular usage and ranges of measurement.

Vertical opening

Vertical opening is one of the few important parameters of trawl net. Other things being the same, the catch is directly related to the mouth area. The number of floats and sinkers in a trawl net is estimated quite often arbitrarily and it some times results in the use of inadequate number or too many of them. The measurement of this parameter of the net solves many of such problems and gives an easy and accurate way of estimating the performance of a trawl net. The basic principles used, at present for this measurement are (i) the hydrostatic pressure difference between the points and (ii) the time interval of an acoustic pulse to travel from the head rope to the foot rope or vice versa.

The differential manometer type self recording mechanical instruments described by de Boer (1959), Hamuro et al (1959) and Nicholls (1964) are too heavy for an average trawl net and the information is obtained only after hauling out the gear. They measure the hydrostatic pressure difference between the foot rope and headrope which corresponds to the vertical mouth opening. In these instruments, two pressure capsules fixed on the head rope anf the foot rope pick up the corresponding hydrostatic pressure and convey them to another bulky unit, where the pressures are made to oppose each other. A pen actuated by a force proportional to the difference in pressure records the same on a moving paper. For reducing the bulkiness and obtaining instantaneous information, an alternative method was described by Leon (1968). The transducer of a low range fish finder is fixed on the head rope keeping its direction vertically downward. The ultrasonic pulses emitted by the transducer, after reflection from the special reflectors on the foot rope reach the transducer. The time interval between these two events communicated to the boat by cables is recorded in the echosounder. For obtaining sufficiently accurate measurements, the pulse length of the ultrasonic beam must be very short compared to the vertical opening. This is technically a difficult problem and in an ordinary fish finder, the pulse length itself covers a distance of 2 to 3 meters and consequently the vertical opening itself is submerged in the starting mark recorded in the instrument. With the help of special echosounders of very short pulses, the vertical opening can be measured fairly accurately. Holme and Mills (1969) described trials taken with the cableless telemeter type depth recorder developed by M/S Furno Electric Co. Japan. This is a recently introduced sophisticated instrument for obtaining informations of the area below the head rope of the trawl net. High frequency pulses are transmitted from a transmitter fixed on the head rope with its transducer directly downwards. Reflections from the fishes, foot rope and sea bottom reach the transducer and after amplification, ultrasonic waves of different frequency modulated by the reflected signals are transmitted to the boat. These signals are received and recorded in an ordinary echosounder whose working is synchronized with that of the one kept under water. M/S Koden Electronics Co Ltd, have also developed a similar equipment named "Net Monitor".

Sivadas has developed a prototype of a vertical net opening meter based almost on the same principle as those of differential types. Instead of recording the information there itself, they are converted into electrical signals and conveyed to the boat by means if electrical cable. The size of the transducer has been brought down to $1/15"^1$ of that of the commercially available type weighing below 2 kg.

Horizontal opening

The horizontal opening is equally important as the vertical opening and is more significant in bottom trawling, because bottom fishes, especially prawns will be usually moving at the bottom, within two meter range and a further widening of the net in the vertical direction is of little use, while increasing it in the horizontal direction will proportionally increase the catch.

The horizontal opening is usually estimated by measuring the warp angle. The warp angle in radians, multiplied by the warp length gives the distance between the otter boards. This is very much approximate as it is based on the assumption that the warps are straight in the water. But the curvature developed in the warps produces a substantial error in this measurement. Further the angle subtended by the warp will be of the order of 10 to 15 degrees and it is difficult to measure it within 2 degree accuracy with the present method which is quite improvised. The corresponding error in the estimated opening comes in the order of 20%. Nicholls (1964) described a potentiometer type "divergence meter" for measuring the warp angle accurately. He has used two potentiometers with telescopic arrangements and graduated their relative variations in horizontal distance. De Boder (1959) described a recording type instrument. By connecting a thin flexible cord between the two otter boards, the opening was measured with the help of a recorder fixed on one of the otter boards. The string is always kept taught by the help of the springs used in the unit. In addition to all the disadvantages of a self recording heavy instrument mentioned earlier, this is very inconvenient to operate. With the help of two transducers of a fish finder fixed on both the otter boards and obtaining pulses transmitted from one transducer to the other Leon (1958) has measured the opening fairly accurately. Nicholls (1964) described a very sophisticated electronic underwater transmitter developed by electronic department of Saundero Roe. The pulses from the transmitter in one of the otter boards is received in the other and is retransmitted to the original one in a different frequency. The time interval is recorded there itself. Nicholls (1964) describes another type also with the two similar type transducers, one for transmission and the other for reception both being connected by electric cables.

Water current

The velocity of ocean currents vary from zero to several knots. The magnitude and direction of currents vary in oceans, sea and back waters. The performance of a fishing net depends very much on the currents and hence the measurement of water current is very important. The modern commercially available current meters generally use a light propeller as the transducer for picking up the magnitude of current. The propeller with the help of a magnetic field produces electrical pulses proportional to its r.p.m. The deviation of an energized coil from a similar one which has been set to the magnetic meridian of the earth, produces a signal giving an indication of the direction of the current with respect to magnetic meridian. Many producers like M/S Kelvin Hughes (London) and M/S Toho Dentan (Japan) have marketed such current meters. A different method using thermistors has been developed recently by Doughlas (1968) for the accurate measurement of very low currents which are not detectable by other usual methods.

Mesh size variation

The meshes of the trawl net undergo stress and strain in operation. The strain produces affects the shape of the meshes and this variation in shape will be different at different portions of the net- These changes in shape affect the normal functioning of the net. Too much elongation of the net reduces the gap and causes high resistance to motion of the net. Further it affects the total length of the net considerably. The study of the mesh size variations helps the designer to foresee the changes the net undergoes while in operation and make necessary modifications.

No instrument is at present available for the measurement. The instrument must be very light and small in size so that its presence in the net should not in any way affect the functioning of the meshes. Sivadas has made a very light inductive type linear transducer to pick up the mesh shape variations. The variations are converted to electrical signals and conveyed to an electronic indicating meter on board the vessel, where the signals are translated to lengths. Several numbers can be used at different portions at a time and simultaneous measurements can be obtained.

Water flow inside and outside trawl net

The comparative variation of the water flow inside the net with respect to that of outside, is a measure for opening of the meshes as well as the resistance offered by the net.

Hamuro and Ishii (1964 a) have made an underwater recording type mechanical instrument for the measurement of the magnitude of the water flow inside the net. Sivadas has made a prototype of a very light inductive type water flow transducer along with a solid state electronic indicating meter for this measurement.

Other parameters

There are many other parameters, such as warp declination, otter board action on ground, ground rope curvature etc. which are however, not as significant as the ones mentioned earlier

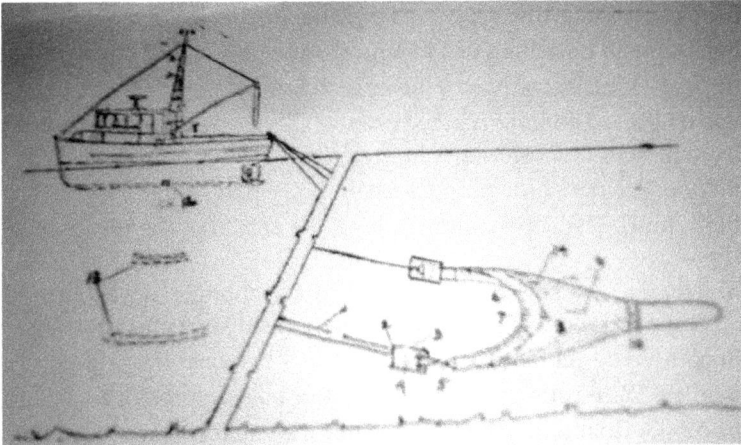

Fig. 141: Positions of the telemetering instruments for the study of the behavior of trawl.

1. Ship istalled tension meter. 2. Tilt meter. 3. Angle of attack meter. 4. Fore and att tilt meter. 5. Under water tension meter. 6. Net depth meter. 7-8. The pressure capsules of the net height meter. 9. Water flow meters. 10. Tension meter for cod end tension. 11. Electric cable. 12. Acoustic transducer. 13. The ultrasonic pulses, 14. Mesh size meter.

Technical terms used in trawl gear technology

1. Assembling the net – Joining the different panels of webbing together and mounting with reinforcing ropes so as to form the net ready for use.

2. Assortment – Bails of machine-made webbing .having specific mesh size

3. Backstrop – A piece of wire rope with high breaking strength used for attaching the otter board with the sweep line and is partially responsible for determining the angle at which the boat will tow.

4. Bar – One of the four sides of a mesh, also called half mesh.

5. Bar cut – Cutting only one leg at each knot.

6. Bar length – The length between two successive knots on a mesh usually used to denote mesh size in mm.

7. Belly – The panel forming the main body part of the trawl net.

8. Belly line - The rope running along the seams of the entire length of the trawl

9. Bolch line – A soft rope to which the webbings are hung and in turn attached to the working ropes, namely head rope and foot rope.

10. Bosom – The center portion of the trawl between the wings.

11. Bridle – Wire rope connecting the otter board to the net.

12. Buoy – A float with conspicuous marl, some time carrying a flag or lamp used to locate the presence of the gear or to identify a ground. In some countries radio-buoys are also in use.

13. Buoyancy – Ability to float in water. Extra buoyancy means the capacity to float an appreciable mass along with the floating object, namely, floats and buoys used on trawl nets.

14. Butterfly – An iron bar with a wide angle attached along with the danleno to the trawl net for achieving vertical spread.

15. Batting – Reducing number of meshes in successive rows to achieve the conical shape of the panel in trawl nets. Achieved by employing different cycles of cutting. While fabricating the panel manually two meshes of the preceding rows are caught up in a single knot in the succeeding row.

16. Beam trawl – Fore-runner of modem otter trawls which used a metallic or wooden beam to maintain the shape of the trawl mouth.

17. Braiding – Mending the net by hand.

18. Cod end – The cylindrical end portion of the trawl net usually made of small meshes of thick double twine where the fish gets accumulated during trawling operation.

19. Cod line – A rope with high breaking strength threaded through the meshes of the lower periphery of the cod end so as to close the cod end.

20. Combination rope – Rope made of natural or synthetic fibers with the strands having reinforced core of galvanized steel wire.

21. Cork line – Old term for head rope.

22. Creasing – Increasing the number of meshes in successive rows by cutting or by braiding.

23. Danleno assembly – Consisting of iron bobbin, swiveled on one end to the butterfly and the other end to the bridle on each side to achieve the horizontal spread of the net.

24. Denier – A unit for measuring the linear density of yarns equal to the weight in grams per 900 m length.

25. Depth of panel – Width of the net expressed in number of meshes running vertically or in number of rows.

26. Double mesh – Meshes made of double twines usually used at junctions of the net and also for the cod ends where excess strain is experienced.

27. Elasticity – Property of a material with which it returns to its original dimension and shape on release of deforming force.

28. Elongation – The extension in the direction of load caused by a tensile force and is expressed as a percentage of the original length of the material.

29. Eye-splice – A small loop spliced at the end of a rope and reinforced with metallic thimble.

30. False belly – Strip of webbing attached to the underside of the lower belly to prevent the net from tearing due to friction on the rough sea bed.

31. Fiber – Basic material used for manufacturing yarns. May be of animal, vegetative or synthetic origin.

32. Fishing gear – Commom term for the tool or tackle employed for catching fish.

33. Flapper – A small piece of webbing shaped to fit in the interior of the trawl and hung from the upper belly intended to prevent the escape of fish from cod end to trawl mouth

34. Figure–Eight–Link – A metal link shaped like a figure eight and designed to join the Kelly's eye.

35. Foot rope – A synthetic or steel wire rope lining the lower bosom and lower wings with sinkers or chain pieces attached to it in order to add negative buoyancy to the lower panel of the net. It is also known as foot line or lead line or ground rope,

36. Fly mesh – Free hanging meshes having only two knots usually made at the edges of webbing for hanging the webbing. Also known as "Dog ear"

37. Float – Materials having extra buoyancy like cork, wood, aluminum, plastic etc. attached to the head rope of the nets to keep the upper portion of the net floating. Accounts for vertical opening of the trawl mouh.

38. "G"-Link assembly – A special clip link with a counter link shaped exactly to slide in, used for locking in of otter boards to the sweep line.

39. Gallows – Strong inverted "V" shaped points on the aft of the trawler one on each board from which the gear is towed.

40. Ground rope – Sometimes an additional rope with rubber rollers and metallic bobbins attached to the foot rope by means of links to save the gear in rocky and corally grounds.

41. Hanging the net – The process of attaching the webbing to the supporting or surrounding ropes. Done as per specific hanging co-efficient for each type of gear and also for different parts of the same gear.

42. Hanging co-efficient – The ratio between stretched length of the webbing and the total length of the rope on which the webbing is hung. Owing to the flexibility to the geometry of the webbing, the hanging co-efficient has two complementing components, the horizontal and vertical hanging co-efficient.

43. Head line elevator – A buoyant plate fitted with floats attached to the head rope for increasing the vertical opening of trawl net.

44. Head rope – A synthetic or steel rope lining the periphery of the upper bosom and upper wings to which floats are attached to gain extra buoyancy for the upper panel of the trawl net. Also known as head line, cork tine or float line.

45. Hemp – A plant, the bark of which serves as a source of hemp fibers widely used in olden days for making twine and ropes.

46. Jib – Triangular piece of webbing attached to both sides of upper and lower edges of the mouth portion of the trawl net, also known as wedge.

47. Jute – A plant widely cultivated in India and Pakistan, a source of vegetable fiber.

48. Kelly's eye – Shackled to the double end of the back strope. It is a combination of two metal rings, one for a shackle and the other for joining the figure of eight link.

49. Kapleon – A synthetic polyamide fiber used in fishing industry.

50. Knot – A tie made by two ends of a twine or a rope etc. to join the ends or a tie made by one end at some parts of its own body to make a loop or over some other object to get it fasten to it.

51. Lacing – Seaming or joining the upper and lower halves of a trawl net along the sides by winding a twine around a few meshes gathered from the edge of each half and fastening at intervals with a jam hitch or stop hitch.

52. Line – A rope usually of a prescribed length for a well defined use.

53. Linen – Yarns made from flax.

54. Manila – Hard fiber taken from the leaf stem of Abaca plant widely used as twines and ropes in fishing industry.

55. Mesh – Inter-spaces of a fixed dimension formed by a sequence of knots.

56. Mesh size – Measured variously as bar length and stretched length, the former being the distance between two successive knots and the latter being the total stretched length between the mid-points of the two furthest knots (Mesh size stretched).

57. Monofilament – A single continuous filament of same synthetic fibers.

58. Multi-filament – A filament yarn having several numbers of individual filaments.

59. Needle, mending – Wooden or metallic tool used in braiding and mending nets.

60. Net – A fishing gear of definite design made from webbing.

6I. Net, knotless – Webbing made on power looms by inter-weaving two twines at junction without making knots.

62. Nylon – A synthetic fiber belonging to polyamide group.

63. Otter board – Large wooden or steel board mostly rectangular or oval in shape rigged to the trawl nets. By virtue of its calculated angle of attack to the sweep lines and the warp it diverges side ways and gives required horizontal spread to the trawl mouth. Also known as otter door, trawl door.

64. Otter trawl – Modern trawl gear rigged with otter boards. A name coined earlier to differentiate from the beam trawl.

65. Ply – Number of yarns in a strand or total number of yarns in a twine. 20/4/3 means twine consisting of 3 strands and each strand consisting of 4 yarns (plys) of 20 counts.

66. Point – A knot with one or two legs along the edge of webbing.

67. Preservation of net – Treating the webbing with specific chemical solution to prevent slippage of knots, decay of vegetable fibers and reaction of sun beams (in case of synthetic fibers).

68. Quarter rope – Ropes attached to the trawl net for hauling the net.

69. Reef knot – Common type of knot used in net making. Also known as square knot or flat knot.

70. Reel – The equipment on which yarn is wound and the equipment used for hauling hook and line gear.

71. Rotting – Deterioration of strength of the net due to action of moulds and bacteria in water. Higher water temperature quickens the rotting.

72. Seam – Laced edge of net.

73. Seam line – Line laced along the seam.

74. Selvedge – The two sides of the netting (length wise).

75. Setting knot – Fastening the knots of webbing by treating with steam or by adhesive preservation.

76. Silk – An animal fiber produced from pupa of silk worm.

77. Sisal – Fiber obtained from leaf of sisal plant.

78. Sling – Splitting rope used in hauling the catch.

79. Splicing – Joining two ends of yarn, twine or rope by inter-weaving of its strands.

80. Spreader – Short wooden or iron piece attached to the wing tip of trawl net to keep the wings vertically stretched.

81. Square – Front portion of the upper panel of the net between head rope and belly, which overhangs the lower portion of the trawl net. It prevents fish from swimming upwards from the trawl mouth. Also known as overhang.

82. Strand – An assembly of a number of yarns kept together by twisting them about its axis. Strands are twisted together to form the twine.

83. Sweep line – Wire rope which connects the trawl wing to otter boards. It sweeps along the sea bottom and hence the name.

84. Strop – A length of wire rope with the ends short spliced.

85. Take up ratio – The ratio between the stretched length of webbing to the length after hanging.

86. Tenacity – Breaking force in terms of fiber or yarn denier expressed in grams per denier.

87. Tensile strength – Breaking force in term of unit area expressed in grams per square mm.

88. Terylene – A synthetic fiber belonging to the polyester fiber group.

89. Thread – Strand or strands of yarn twisted into a fine line of twine.

90. Tickler chain – Iron chain attached ahead of the foot rope of a trawl to frighten the bottom fish and to induce them to enter into the net.

91. Trawl net – Complete net with all sections joined together having a bag shape and dragged by a vessel.

92. Twine – An aggregate of fibers or yarns twisted.

93. Twist – Turns about the axis of fiber, yarn or twine to keep them together. Described as soft, medium and hard twisted according to the twist per unit length.

94. Trawl winch – The mechanism for paying out and heaving the net and its accessories.

95. Warp – The line by which the vessel tows or drags the trawl net. Also known as towing cable, towing rope.

96. Webbing – A sheet of netting used for fabricating fishing nets.

97. Wings – The two end sections of the trawl net with a broad base and tapering to the ends.

98. Vertical opening – The distance between the mid-points of the upper and lower bosoms of a trawl net while in operation. Achieved by adding floats along the head rope and sinker along the foot rope.

99. Yarn – A number of fibers or filaments twisted together around a single axis to form a continuous line.

100. Yarn count – Unit of expressing linear density of yarn.

Fish Behavior to Approaching Trawl Gear

A moving trawl provides a set of optical and acoustic stimuli, which influence fish in many ways. Fish possess many types of sensory receptors that may detect the presence or approach of the trawl gear. Their environment may change from light to dark, which determine visible range and how far the fish can see the stimulus of the approaching trawl. Water temperature is a factor affecting speed and endurance in out-maneuvering the trawl. Direct observations of fish reactions to towed gears allow to predict in a general way what fish are likely to do when approached by a particular trawl gear,

The visual stimulus provided by a trawl gear depends on the color and mesh size of the netting panels, and on the various bridles and ground gear employed, The audible stimulus depends on the roughness of the sea bottom, but is also greatly affected by the mounting of head line floats and the presence of bobbins, chains etc. in the ground gear. Seine nets or lightly rigged trawls are almost silent on a sandy bed; whereas heavy trawl ground gears can generate loud noises on sand or hard sea beds.

The reaction of some commercially important species of fish to two contrasting types of bottom trawl was observed from an underwater vehicle. Both nets appeared visually similar underwater, but the heavily rigged bobbin ground line of one was easily seen from the wing ends, whilst the light grass footrope of the other was not visible. The dragging noises of bobbins trawl could be heard 10 meters ahead of the wing ends, whereas the light trawl was virtually silent. Vessel's noise could be heard at each net. All fish swam directly to the nets in the warp, otter board and bridle areas. Haddock *(Melanogrammus aeglefimts)*, Saithe *(Pollachms virens)* and Mackerel *(Scomber scombrus)* swam straight into the lightly rigged net, but most turned ahead of the heavily rigged bobbin foot rope. Haddock and saithe approached the gears at average swimming speeds calculated at 0.2 to 0.4 meter per second and 0.7 to 0.85 meter per second respectively. All species were seen to feed on small sand eels in the mouth of the trawl. Many haddock easily escaped over 4.6 meter high head line of the bobbin trawl, but none escaped from the lightly rigged net.

The sand clouds from the otter boards were similar in size and shape and the clouds passed one to two meter outside the wings of both trawls. When the vessel was not towing a straight course, due to variation in wind direction and tide, the inner edges of the clouds alternated between two meter and three meter on either side of the wing tips.

Both gears had good contact with the flat sea bed from the otter boards to the bosom ground line. Both nets showed a good shape and only small areas of loose netting were seen. The rear part of the belly panel of the lightly rigged net however, was slack and fishes were frequently seen pressed on to these meshes. Some times the cod ends would twist and turn slowly when well filled with fish. After two or three revolutions in one direction, the entrance to the cod end was temporarily blocked preventing any fish from passing back into the cod end. After approximately half a minute, the cod end slowly unwound and the fish passed into the cod end. This effect was not seen when the cod end was almost empty. Previous direct observations have shown that there is a great deal of turbulence in the cod end which probably accounts for this slow spinning effect.

With the underwater visibility of 5 to 15 meter in diver's eye, the netting panels in the front sections appeared grayish in color and visually similar when seen from two to three meters above the seabed. Looking back into the mouth of both nets from the wing area, the netting was virtually invisible against the grey sand cloud background. The light grass foot rope was very faint in comparison to the heavy black rubber bunts and bobbin wheels of the heavy trawl. Both could be seen from the wing ends, but not from 5 meter ahead of this position. Sand rose higher from the heavy bobbin ground line than from the light foot rope. Neither net could be seen from a position half way to the otter boards.

The vessel's engine noise was clearly heard at all times between the otter boards and the net. The sounds heard were due to chains, shackles and bobbins as they moved across the flat sandy sea bed.

Fisb reactions to the lightly rigged approaching trawl

Viewed at the otter board, haddock and saithe were generally seen first both as small groups (10 to 20 fish) and large schools (more than 100 fish) close to the sea bed (1 to 2 meter), slowly swimming towards the net. These species were never seen to swim in the towing direction in this area or to swim close to the otter boards. They swam in a region equidistant between the warps and sweeps. Observations at the wing end showed that the haddock and saithe continued on this path, they then swam straight back over the bosom ground line into the net. or turned in front of the foot rope, close to the sea bed, to remain with the gear. Observing still further back, through the wing of the net, a group of 20 to30 haddock (34-45 cm) were seen in the path of the approaching net feeding on sand eels *(Ammodytes marinus)*. As the foot rope approached them they suddenly took up station with the net by swimming steadily forward close to the sea bed just ahead of foot rope. Similarly saithe were seen to feed on small sand eels both on the bottom and up to 1.5 meter above it and up to 5 meter ahead of the bosom ground line. A few saithe were hit by the ground rope while feeding on the bottom, one fish appeared stunned and another was run over by the light foot rope. Normally, saithe avoided the sand clouds which occasionally came inside the wing ends of the net, but they were diving into the low sand clouds thrown up by the ground line to feed on small sand eels. Those haddock and saithe that swam straight into the net, turned 5 to 6 meter behind the foot rope to swim forward with the net. At a towing speed of 1.5 m per second, most saithe swam close to the belly meshes to a position near the foot rope. Only the large haddock (40-50 cm) could sustain this effort and these only for a short time. On one occasion a large school of haddock

(400-500 fish) was observed to turn between the wing ends and form a narrow column of fish in the center of the net path.

Haddock swimming ahead of the ground line slowly lost ground and eventually, after no more than two minutes, were in a position 1-2 meter ahead of the foot rope bosom. These fish then showed a "kick and glide" swimming action to stay in front of the gear at 1.5 meter per second speed. No haddock maintained this speed for more than two and half minutes before they tired. They then rose high off the sea bed, turned and slowly swam into the top part of the net. None was seen to escape over the head line.

Saithe (30-45 cm) swam for larger periods ahead of the net mouth (approximately 15 minutes at 1.5 meter per second) and then rose clear of the foot rope and passed back into the net.

Further back in the tunnel of the net many haddock and saithe turned and oriented themselves to swim forward again, but more slowly than the net, so they steadily lost ground and dropped back into the cod end. Stomach analysis of the haddock and saithe after completion of these hauls confirmed the presence of undigested sand eels of 11-13 cm.

One group of mackerel (27-45 cm) was seen swimming 3-5 meter outside and ahead of the port otter board. This school (approximately 100 fish) swam at right angles to the gear and into the path of the trawl both under and over the port trawl warp, which was approximately one meter of the ground. They then swam directly away from the warp, close to the sea bed, and when equidistant between the trawl doors, turned and swam towards the advancing net. From near the starboard sweep wire, another large schools (100 to 200 fish) was seen 2 to 3 meter inside and ahead of starboard door. These fish turned and passed out under the starboard warp and swam away from the path of the gear. On both these occasions, the mackerel swam fast (2-3 meter per second) using steady tail beat movements.

Small groups (5-10 fish) were also seen between the bridles, mixed with larger schools of haddock and saithe, swimming directly towards the net, close to the sea bed. Some swam straight back into the net, but others turned four of five meters ahead of the foot rope, rose up from the bottom and formed a separate school above the haddock and saithe. This school moved around the mouth and inside the net and finally returned to a position, near the center of the net mouth, close to the top panel, whence after a few minutes, it moved forward out of the net. These mackerel swam past and through schools of saithe and haddock swimming ahead of the foot rope. Then they increased their speed again and swam away between the wings, out of the gear. These mackerel did not return whilst these observations are made. Of the few mackerel recovered from this net, most were meshed in the batings.

A mackerel of 40 cm swimming among a school of saithe above the foot rope was seen to swim quickly forward to catch a sand eel, lose it, dash forward again to take another, and then steadily swim forward out of the trawl and escape. Towing speed at that time was 1.5 meter per second,

Fish reactions to heavily rigged bobbin trawl

In the otter board area haddock (25-50 cm) and saithe (27-46 cm) behaved in a similar manner in small (5-10 fish) and large groups (100-200 fish). They swam slowly towards the approaching gear, equidistant between the boards. On four occasions just ahead of the boards, 20-30 saithe were seen close to the sea bed and feeding on the bottom. When the otter boards passed they swam slowly towards the net, between two sand clouds. These species never swam in the towing direction in the otter boards or bridle areas. Near the net itself, haddock and saithe again behaved similarly. Once, level with the wing ends, most saithe turned to swim with the net; whilst a few swam out under the wing line, most kept well clear of the wings and close to the sea bed. During one observation period, a large number of saithe (100 fish) swam back towards the net and passed out over the port sweep ahead of the bridles and into this corridor of clear water, swimming with the gear between the sand cloud and the wing of the net. The sand cloud was 3 to 4 meter outside the port wing end, because the towing vessel had altered the course slightly to starboard. The fish in the corridor had tired and dropped back outside the port bridle and were being held between the sand cloud and the net. During another period, when the vessel was manoeuvring, 50 to 60 haddock swam back close to port sand cloud and entered the clear water corridor between the bridles and the sand cloud- Only a few were seen to pass under the wing line into the corridor. Haddock were never seen to enter a sand cloud, but on one occasion a few swam up over it, while others escaped by passing back along the outside of the net in the clear passage between the net and sand cloud. However, most haddock turned to swim with the gear 5 to 6 meter ahead of the bobbins, but many turned as far as 5 meter ahead of the wing ends. On six occasions, groups of haddock and saithe met the net at the height of the wing ends (wing end height 2.5 meters). All these fish turned forward and then slowly descended on to the sea bed between the wings to keep the station with the net. They spread out low across the path of the net, and other fish coming back from ahead of the gear turned and joined the front of the main school to form a narrow ribbon or column formation, equidistant between the wings. No stacking of haddock was seen to occur at this position, even when an estimated 400-500 fish were swimming between the wings ahead of the bobbins. Saithe in this position also kept close to the sea bed, but when large numbers filled the area between the wings, they stacked up to a height of 1 to 2 meter, just above the haddock. At 1.5 meter per second, saithe swam for approximately 15 minutes before dropping back, whereas haddock swam for only 2 to 3 minutes, before they rose up out of the school to enter the net 3 to 4 meter above the bobbins. On some occasions, when the vehicle was outside the port quarter and the visible range was 10 to 15 meter, haddock were seen to rise from the sea bed as far ahead as the wing ends and swim upward to an estimated height of 8 to 10 meter, before they turned back to escape over the head line (head line height 4.6 meter). Haddock escaped over both the port and starboard wings as well as the head line center, and it was estimated during one haul. 30 to 40 percent of the haddock catch was lost (approximately 3000 to 4000 fish).

Inside the net, haddock swam forwards using a fast kick and glide action, but slowly lost ground and dropped back along the funnel. As with light rigged net, many saithe turned again 5 or 6 meter inside the net itself, descended close to the belly panel and then moved forward to the bobbin foot rope, using steady tail beat movements, until they again tired and dropped back towards the cod end.

A "flapper" is fitted into the funnel of the heavily rigged trawl to prevent fish escaping from the cod end in the event of the net stopping during hauling or coming fast.

At the flapper, all fish panicked as the surrounding netting became more constricted. They maintained the fast tail beat movements and swam with an erratic kick and glide action to try to hold a position ahead of the flapper. Large and small haddock and saithe were squeezed and bumped as they slid and scraped along the flapper panel before entering the cod end- There was physical damage to all fish passing through the area of the net.

In another experiment, 400 to 500 saithe were swimming in the mouth of the net, when the vessel was instructed to "knock out" and haul the gear onboard in the normal manner. On "knock out", the gear immediately slowed from 1.5 to about 0.25 meter per second, and the saithe continued to swim forward at 1.5 meter per second. The vessel being a side trawler, towed the gear to starboard and the port board cut across the original path of the trawl. This caused the port side sand cloud to cross the path of the saithe. They stopped swimming, milled around and slowly dispersed. As the sand cloud cleared after 30 seconds, many of the fish still in this area reformed the school and swam forward again but away from the direction of the obliquely moving gear as it was being hauled back to the vessel at approximately one meter per second. It was estimated that a large proportion of the saithe was lost due to this side trawling "knock out" manoeuvre.

It is of interest that each time the divers were alerted (27 times), basing on the cchosounder recordings, fish arrived in notable numbers at the mouth of the net. The clock on the video frame and the time of the alert recorded on the audio track on the recorder gave the time taken for the fish below the vessel to reach the net (towing speed 1.5 meter per second) from which average swimming speeds were calculated. Haddock swam towards the trawl at speeds 0.2 and 0.4 meter per second, whereas saithe swam at speeds between 0.7 and 0.85 meter per second.

In day light most fish either moved steadily away from the vessel in the direction of the trawl or remained in position until movement was triggered by the approaching gear.

Haddock rising and turning either at the wing end area or well ahead of the bobbins have more time to see the gear and escape over the approaching headline than fish swimming very close to or behind the foot rope. No haddock escaped over the head line of the lightly rigged net, yet many easily escaped over that of the bobbin trawl. In darkness or in turbid water their reaction might be different because the components of a fishing gear will be seen at much shorter distances.

Contrast perception in cod is as good as or considerably better than in humans at most of the light levels studied.

The noise from the bobbin trawl generated by the chains, shackles and bobbin spacers was very pronounced in comparison to relatively silent light rigged trawl.

18 | Engineering Performance of a Trawl Gear

Flume tank of White Fish Authority (WFA) for development of more efficient trawl gear.

Over the years fishermen have designed nets which are surprisingly efficient considering the difficulties they have had to face. Chief among these is that of designing and working in the dark. Fishermen have had to imagine or sense the shape their net would assume in the water, just how it would perform and how fish would react to it. As a result, the development of new or improved fishing gear has been a slow evolutionary process. The introduction of modern instrumentation techniques to measure the engineering performance of a trawl has considerably speeded up this process. But it is still time consuming and very expensive to test full-scale gear at sea.

However, because of such work with instrumented trawls, a large body of mathematical knowledge has been built up about the engineering behavior of trawls and their rigging, which has enabled model trawls to be designed and then tested to check out the performance of any new design concept before engaging in full-scale trials at sea.

The WFA flume tank enables large scale models of prototype trawls to be subjected to a wide variety of conditions, and their response to these modifications measured, recorded and analyzed in controlled experiments.

What is important is that this work can be done at a fraction of the cost of trying out a full size net, and a great deal more information can be obtained on geometry and general performance than is possible at sea. This does not mean that the procedure in which a new trawl is tested at sea with full test instrumentation and progressively modified until the required net parameters are correctly set is no longer necessary, but it does mean that it can be considerably reduced.

Development work done at the flume tank can be broadly divided into two parts; the first is made up of investigations done by gear manufacturers, fishing companies and other commercial organizations; and the second is the WFA's own program of development.

Many net makers and fishing companies in the UK and throughout Europe use the WEA flume tank to test new trawls and also to investigate hydrodynamic problems related to existing gear. Component manufacturers are also able to test the smaller items which make up a trawl, such as, floats and bobbins, often at full-scale.

In some cases companies build their own models and book time in the tank to perform their experiments with WFA assistance. On other occasions the WFA takes a more active part, building or advising on the construction of models, designing and performing experiments, analyzing the results and preparing a full report. In both cases the work is done as a part of the WFA's technical consultancy service and a charge is made.

The WFA's own development program is also aimed at developing new trawls up to the point where they can be introduced on commercial fishing vessels The trawls may be WFA-designed, like the large Balta demersal trawl or gear designed by the Marine Laboratory, Aberdeen. In the latter case the WFA assists in development by performing model tests in the flume tank, and later by trying the full-scale gear on commercial trawlers.

Work is also done on improving existing trawls, with relatively simple changes often producing quite marked improvements in performance. For example one particular demersal trawl was reported to be lifting off the bottom instead of fishing hard down. When a model of the trawl was made and tested in the flume tank, it soon became apparent that to achieve good contact with the sea bed the width of the lower wings had to be increased. In a very short space of time this modification was effected and the model successfully re-tested. The full-scale gear was subsequently modified in a similar manner and the trawl is now being used commercially and achieving good catch rates. This demonstrates very forcibly how the flume tank can be used to solve, in a few hours, a problem which would probably take weeks or even months to solve and validate at sea.

The flume tank

In the flume tank, which is the largest of its kind in the world, models of trawls can be demonstrated and tested. As a result it is possible not only to show fishermen how they can make adjustments and modifications to their gear for improved earning efficiency, but also to provide net manufacturers and others with a simple and economical method of testing new and improved fishing gear.

All types of gear on which the effect of water flow is critical can be tested in the tank, and at scales much larger than possible in other facilities. Consequently, problems associated with the correct scaling of twine, mesh sizes and so on are considerably reduced and much closer simulation of the behavior of full-scale gear is achieved.

Measurements made on model trawls in the flume tank are matched with those taken on the full-scale gear at sea using full test instrumentation. These validation procedures ensure that when a fishermen modifies a model in the tank and achieves better performance, the same modification will produce a similar improvement on the full-scale gear he uses at sea.

By developing new and improved trawls and in training fishermen to make more effective use of them the WEA, with its fisheries training center and flume tank, is preparing the way for a more efficient industry in the future.

Description and operation

Structurally it is a large re-inforced concrete tank 31 meter long by 5 meter wide by 5 meter deep, built at ground level. It is divided horizontally into two chambers which are inter-connected at each end to enable the water to circulate.

Demonstration and testing of the model nets takes place in the center portion of the tank upper chamber. This area, which is the tank working section, measures 11 meter long by 5 meter wide by 2.5 meter deep. The lower chamber of the tank forms a return passage through which the water circulates (Fig. 142).

Fig. 142: The net seen through the observation windows in the side of the flume tank

Total water capacity of the tank is 700 cubic meter (approximately 700 tonnes). Mains water and not sea water is used and an additive is used to inhibit corrosion. A filtration system is installed capable of filtering the complete contents of the tank in 24 hours.

The water is circulated around the tank by four impellers driven by an electro-hydraulic system. The impellers are 1.2 meter in diameter and are positioned inline across the down stream end of the tank. At their rated maximum speed of 200 revolution per minute they deliver a total of 1136 cubic meter of water per minute.

For testing model trawls water speed through the tank can be varied from 0 to 1.0 meter per second (0-2 knots) which, at a scale of one-fifth, represents a maximum full-scale water speed of 4.5 knots. The water speed can be increased to a maximum of 1.5 meter per second to permit the full-scale testing of small components such as, floats, kites and transducer housings.

Water flow around the tank is controlled by various suction screens, deflectors, cascade bends, wave traps and flow straightening screens to obtain as nerar as possible uniform water speed through the working section.

A large moving conveyor belt forms the floor of the working section. The conveyor stimulates the ground friction which a demersal trawl experiences when towed across the sea bed. In addition, the movement of the belt eliminates the velocity difference which would otherwise cause eddies close to the lower surface.

The conveyor is driven by its own electro-hydraulic system and can be run at the same speed as the water, or at different speeds to stimulate sea bed currents.

The model trawls are not towed through the tank, but remain stationary when under test, the water flowing through and around them. They are worked from a platform above the tank, the warps being attached to towing points, adjustable in height and width, upstream of the working section.

Observation of the models in the tank is made through targe windows in one side of the tank and from a motorized trolley which runs on rails along the top of the tank. The windows are 1.5 meter high and I.I meter wide and extend for an overall distance of 11 meter. The glass is 38 mm thick and especially cast for the purpose.

Banks of flood lights at the back and top of the tank are used to assist photography or when the ambient light level is low and a glass bottomed box, which overcomes surface diffraction, is used when observing the trawls from above.

Accurate measurements of vertical net dimensions, wing end spread, door spread and so on are obtained with special optical instruments outside the tank and a series of calibrated lines on the conveyor belt and the back of the tank enable rough checks to be made for non-scientific works.

A console in front of the observation windows houses all the controls, gauges, stop-start burtons and other items of equipment required for the operation of the tank. Separate warp tension meters are installed and an intercom system, with public address and talk facilities, is also provided to enable the fume superintendent to communicate with personnel on the working platform on top of the tank.

Models

A comprehensive range of model fishing gear is available for demonstration and testing in the flume tank and new ones are continually being added. They range from small inshore trawls used by vessels of a few hundred horse power, to the distant water trawls towed by vessels of 2000 horse power or more.

The models are scaled down representations which retain the essential features of the original gear. Their use is an important means of gaining under-standing of the performance of the full-scale trawls and communicating this understanding to the industry.

The size of the flume tank enables models to be built to larger scales than are possible in other facilities. Typical of the scales adopted for fully-rigged gear are one-eighth for inshore demersal trawls and one-twenty-fifth for large distant water pelagic trawls. When the nets only are modeled much larger scales are possible.

Before a model is is built, detailed information on the construction of the full-scale gear is obtained from the manufacturer, Dimensions of all net sections are taken, together with the number of meshes, types and sizes of twine used, hanging ratios and so on. This confidential information is not, of course, released to other individuals or companies.

The theory used in scaling down full-sized gear to model scale is based upon that employed by the Marine Laboratory, Aberdeen and by ISTPM in France. Basically, the rules of this theory require that all dimensions of length are reduced to the chosen scale, weight and buoyancy are reduced to the cube of the scale and towing speed is reduced by the square root of the scale.

Naturally it is not possible or even desirable, to scale down everything to precise accuracy, but the essential forces produced by the various components of a trawl are scaled correctly and it is these that matter. For example, although it is rare for a model to correspond mesh for mesh with the original, it is weight, drag and spreading forces are scaled correctly.

Construction of a model trawl is a lengthy process and requires close attention to detail if it is to have the same shape in water as the full-scale gear. The time taken to construct a fully rigged trawl varies from 250 to350 man hours, depending on the complexity of the full-scale gear.

Most of the models are made by the flume tank stuff. Not only do they make the nets, but the bobbins, floats, trawl doors, bridles and many of the smaller components as well. In some cases net manufacturers and others construct their own models for testing in the tank, advice on scaling being provided by the WFA.

Twisted nylon twine is used for the nets, with sections normally made in double twine (cod ends and extension pieces) being made by lacing together two sections of single netting. The trawl doors are made from brass, stainless steel or oak (depending on the type of door) and ballast is added to obtain the correct scaled weight and center of gravity. Bobbins are made from a glass fiber and cork mixture and floats from balsa wood coated with varnish.

Once a model trawl has ben made, reference is made to trials data on the engineering performance of the full-scale gear at sea. From this information the model can be correctly set up in the flume tank and validation procedures carried out. The aim of validation is to establish that the model is indeed an acceptable representation of the original. This is done by comparing the behavior of the model with a selected set of parameters of the full-scale gear.

Only when a model has been validated it is ready to be used as a substitute for the full-scale gear for the purposes of demonstration and experimentation in the flume tank.

Technical specification of the flume tank

Dimensions

Tank overall dimensions (internal)

Length - 31 meter; Width – 5 meter; Depth – 5 meter

Working section dimensions (internal)

Length – 11 meter; Width – 5 meter; Depth – 2.5 meter

Observation windows

Length – 11 meter overall (11 panes); Height – 1.5 meter; Thickness – 38 mm

Water capacity – 700 cubic meter

Velocity of water flow - For testing trawls – 0 to 1.0 meter per second (variable)

Maximum – 1.5 meter per second

Flow generating pump sets

Number and type – Four, independent, electro-hydraulic driven

Total capacity – 19 cubic meter per second

Speed – 200 revolution per minute

Impellers : Type – Four-blade, fixed pitch; Diameter –1.2 meter

Electro-hydraulic system : Each unit comprises of –

1. 93 kW electric motor, driving
2. Dowty Dowmatic hydraulic pump unit
3. Staffa B80 motor directly coupled to impeller shaft

Conveyor– Length – 11 meter; Width – 5 meter, Speed – 0.1 to 1.5 meter per second

Drive – Toothed belt from hydraulic motor;

Hydraulic motor – Staffa B30

Power unit – Electro-hydraulic power pack

Water filtration equipment

Filter – Royles high flow disposable cartridge

Filtration – 100 to 1 micron

Flow rate – 9 liters per second

Pump – Electric motor driven, centrifugal

Control console :

Controls – Water speed; Conveyor speed, Impeller pump sets start/stop

Conveyor start/stop; Tank filter pump start/stop; Submerged lighting;

Temperature selector

Instruments – Mean water speed; Conveyor speed; Motor running lamps (6);

Impeller shaft revolution per minute (4); Hydraulic boost pressure (4)

Sample main hydraulic pressure, Hydraulic oil temperature; Oil level alarm; Oil temperature alarm

Special equipment – Motorized observation trolly on tank top;

Glass-bottomed box for viewing trawls from above;

Flood lights for illuminating nets in water;

Water speed indicator calibrated to full-scale knots (adjustable)

Warp tension meters;

Instruments for tank calibration;

Instruments for measuring trawl gear in tank;

Intercom system; Hoist;

Workshop and net loft with full range of model making tools.

19 | Use and operation of trawl Gear

Trawler– the vessel that operate trawl gears

By the trawler it is meant that pulls a bag net through the water. There are many types of trawlers varying in size from open boats powered by outboard engines to huge factory ships which can fish in the most distant waters. So the word "Trawler" is a very general term and for the sake of precision it must be qualified by the type of trawling engaged in and size of the vessel.

Trawler requirements

There are two requirements that are commn to all trawlers. One is the need for traction or towing power and the other is the need for a winch or mechanical hauling system. In order to have a good towing power, a trawler should have a reasonable draft or displacement and should have a large slow-turning propeller. Main engine horse power will not itself make a vessel a good trawler. If the engine revolution per minute (rpm) is high and if there is not a big reduction gear is fittd, then much of the power will be wasted by propeller cavitation Most trawlers in the 100 HP to 600 HP bracket are fitted with reduction gears that will give them a propeller speed of no more than 400 rpm at maximum engine speed. On larger trawlers the propeller speeds are much less. This is an important point to bear in mind when procuring a trawler engine. It is possible to get a cheaper unit of the same indicated horse power if one buys a high speed engine, but when one adds the cost of the large reduction gear necessary to obtain optimum propeller revolutions, then one may find that there is little price difference between the complete installation and one comprised of a medium or slow—speed engine with a smaller reduction gear.

Propeller nozzles

Many trawlers are fitted with propeller nozzles to give them more towing power. The relative advantages of these nozzles are debatable. Certainly they do increase the thrust from the screw, especially when towing with a warp on each side as on a stern trawler. However, they are expensive, they require skill in designing and fitting and in most cases (vessels under 150 tons) they reduce maximum speed. The effect of the extra thrust is good on stern trawlers, but on side trawlers which often must use a lot of helm to counter wind and tide, the effect is minimal

There are two types of nozzles; fixed nozzles and steering nozzles. The steering type acts both as a rudder and a thrust improver. They are not so favored for fishing vessels. While they certainly increase maneouverability, they sometimes make more difficult to hold a steady course. Fixed nozzles require a conventional rudder fitting. An indirect benefit of a nozzle is that it provides some protection to the propeller when shooting the gear. However, a net can foul a nozzle propeller just as easy as it can foul an open one. The webbing is simply sucked in at the fore end of the nozzle.

Winches

The winch or hauling mechanism is the next important requirement for a trawler. With winches one should always make sure that the model ordered is adequate to the work that must be performed. Hauling or pulling power is usually given in tons, pounds or kilograms per speed (feet or meters per minute). This specification should be clearly indicated as for bare drum, half drum or full drum. A simple general term such as 5 ton winch or 2 ton winch is quite meaningless for anyone wishing to purchase a specific size of winch

Most of the early stern trawlers had their winch positioned just aft of the forward-sited wheelhouse. On later vessels there use was a preference for split winches rather than single two drum winch. Split winches are better in most cases, but they are expensive and have to be hydraulic or electric drive. The use split winches makes it possible to have the warp go straight from the winch drum to the gallows or gantry. At least one right angle bend can be avoided with a single two drum winch if it is positioned fore-and-aft with the warps leading out in opposite directions. With that arrangement however, one warp leads in to the top of the drum and the other to the foot. This can be avoided by using a parallel drum winch.

In many artisanal fisheries the world over, there is still a lack of simple mechanical expertise as is needed when construction power take-offs and winch drive mechanisms. A clutch should be fitted at the power take off point.

From the power take-off point, belt drive is preferable to chain drive in most cases as it has more flexibility and by means of a jockey pulley can be adjusted for different loads. Vessels with hydraulic installations require only a Vee belt drive to the pump. This must be geared to produce the required rpm speeds. In many situations hydraulic drive may prove to be cheaper than mechanical drive in the long run, provided that expertise and spare parts are available locally.

There is a limit to the number of items that can be driven from the forward end of a marine propulsion engine. The alternative which is now gaining favor in most countries is to have a powerful auxiliary engine which is used to drive winch, pumps, generator and other deck gear.

Otter doors

The use of otter boards to open the trawl net mouth horizontally in preference to a beam, was introduced over hundred years ago. Otter trawling developed rapidly after 1920 and by 1950 it was probably the single most productive method of fishing. In the last fifty years there have been further significant developments in otter trawling. Although otter trawlers are classified into three main categories : side trawlers, stern trawlwes and double rigged trawlers; there are actually a large number of variations in the gear used. There are shrimp trawls, prawn trawls, combination trawls, wing trawls, bobbin trawls, herring trawls, semi-pelagic trawls and mid-water trawls. Most trawlers use one type of net for the greater part, if not for all of the year. There are a number of trawlers (small minority, world-wide) which may switch often from one type of the net to another.

All but the smallest of otter trawlers have gallows, gantries or derricks to handle the heavy otter boards. The net hauling system varies greatly depending on the size of the vessel and the type of the trawl used. Light wing trawls and mid-water trawls may be hauled in by power blocks or net drums. Heavy bobbin trawls may be lifted aboard with quarter ropes or Gilson wires. On stern trawlers and side trawlers with German "bang-bang" gear, they are winched aboard after the cables.

Trawl gallows

Most trawl gallows are of the inverted "U" shape, but there are a variety of designs that may be used. These include tripod, bipod, single pillar and gantry arrangements. When designing or selecting trawl gallows one should look for the arrangement that will be high enough to accommodate the largest boards to be used, strong enough to take the maximum anticipated loads, and of a design which will cause minimum interference to other activities on board.

Warp arrangements

Trawl warps must run out from the winch to the gallows blocks or gantry blocks. The position of warp leads on deck, and the "run" of the wires, are important factors in trawler deck layout. A poorly designed arrangement can cause annoying delays when shooting and hauling of trawl and may result in excessive wear on the gear or unnecessary loads on the winch.

From the winch itself a reasonable lead-in length is required to permit easy operation of the guide-on gear. The warp must lead in at right angles to the winch drums. From the first pair of deck leads, the warps are led to the gallows blocks by an arrangement suitable to the type of vessel and method of operation. On side trawlers they usually run to the rail and down the side deck. On stem trawlers they may lead directly to a roller below the gantry.

To prevent accidents and avoid snagging shackles on the vessel or gear, it is a normal practice to have the warps run above head level. Mechanically driven winches must be situated at a convenient drive point. Hydraulic winches can be placed almost anywhere and more imagination can be used in siting them so that a convenient warp arrangement may be made.

To reduce wear on the warps and unnecessary strain on the winch, one should avoid having many bends in the warp run. Sharp bends of 90 degrees or more should be kept to the minimum possible.

Small otter trawls

Structure of fishing gears

Although the sizes of individual parts and mesh size of the net vary according to individual, the trawl net is basically composed of the following parts :

1. Towing rope : wire rope of 8-9 mm diameter; length is 3-4 times the water depth, thus 300 to 400 meter.

2. Otter board : made of wood or resin; size 60 cm width x 125 cm length

3. Rope : nylon, 35 mm diameter

4. Supporting pole : iron bar

5. Wing net: nylon 12 ply

6. Ceiling net: nylon 12 ply

7. Belly net: nylon 12 ply

8. Bag net: nylon 12 ply

9. Float: made of foam plastics

10. Sinker : made of iron or ceramic

Construction of net

	Thickness	Mesh stretched	Width	Length	No.(sheets)
Wing net (a)	12	28mm	100	100	2
(5) Wing net (b)	12	28mm	100	100	2
Wing net (c)	12	28mm	100	100	2
(6) Ceiling net	12	23mm	200-100	200	1
(7) Belly net (a)	12	28mm	100	300	2
Side belly net (b)	12	28mm	100	300	2
Bag bet (a)	12	20mm	100	100	1
(8) Fish catching section (b)	12	20mm	100	200	2
Patch ©	12	20mm	100	200	2

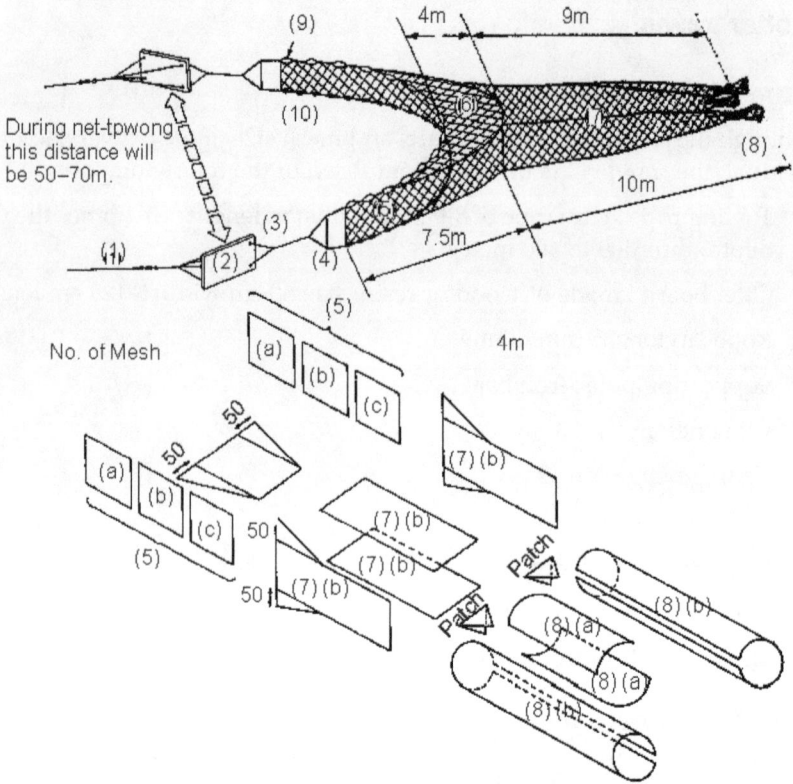

Fig. 143: Diagram of construction details of small otter trawl

Features of the net

The lower part of the belly net and ground rope are connected by a thin rope (shown by the arrow) as shown in the figure and can be easily separated (Fig. 144). The same ground rope is used for 2 to 3 different types of net with different finishes

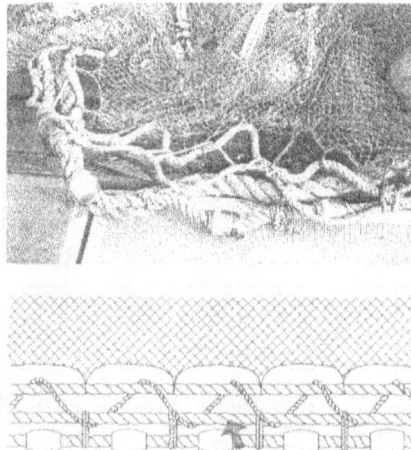

Fig. 144: Showing the ground rope used in the trawl net.

Boat type operating the trawl net

Hull length overall – 14.04 meter; Breadth overall – 3.60 meter; Depth overall at mid-ship – 1.63 meter; Hull weight – 4.42 tons (FRP); Full load displacement – 9.54 tons

Operational method

1. Net casting is to be started at a slow speed cruise. When net casting proceeds until the end of the wing net is reached, the otter boards are placed in position with chains and then by extending the warp, the trawling state is created.

2. Trawling speed (speed against water) is 3-4 knots for net catching large fishes and about 2 knots in the case of small shrimps or prawns. In either case, trawling is performed from up to down according to the tidal current.

3. Trawling time is 40 to 50 minutes per operation and 6 to 7 hauls are performed in a day.

4. Propulsion is stopped for net hauling and the warp as well as the net are rolled up by means of a hydraulic powered net roller while the boat is pulled back. The net roller is quadruplex type and the warp (wire part) is rolled up into the two side reels. When te otter boards are pulled up to the deck, they are hung on the bitt at the stern after the chain is removed. Then the warp and net are rolled up into the two centra! reels.

5. The hydraulic roller is stopped and then the bag net is hauled up and placed on the stem deck by hand. In cases where the catch is large, one should attach a rope to the bag-net part and haul it up using a derrick.

6. The rope tying up the bag end is untied and the catch is dumped out on the deck, After throwing sea weeds and large debris caught caught inside the net into the sea, the catch is stored in the fish holds.

7. The next net casting is done.

8. While the rudder is placed on automatic pilot during net-towing, catches inside the fish hold is sorted.

9. Second net hauling atarts.

Trawl design in small scale mechanized fishing sector

Bottom trawling is known to be a very effective fishing method for exploiting shrimp and demersal fishery resources. Several changes have taken place in the concept of design, construction and operation of trawls over the year. According to the local conditions and availability of fish schools, modifications in design and operation are continuing. Data regarding the designs of trawl nets used in small-scale mechanized sector of Andhra Pradesh, India are out dated and the design

Fig. 145: Operational method of trawl net from a stern trawler.

details, technical and operational aspects and comprehensive modifications in commercial trawls now being operated are described below.

Three categories of trawlers are operated in the east coast. They were classified basing on the length overall (LOA) and horse power, namely, Sona boats, mini-trawlers and large trawlers.

Sona boats of 13.1 meter LOA and powered with 102 hp were mechanized for voyage fishing, to exploit fishery resources up to 100 meter depth.

Sixteen meter mini trawlers with 145 to 180 hp engine operate two demersal trawl gear simultaneously from two out trigger booms at a depth 50 to 150 meter. Multi-day fishing trips (10 to 21 days) were undertaken by these trawlers.

Large trawlers of 21 to 28 meter LOA have engines ranging from 350 to 624 hp. They operate four trawl nets simultaneously, two nets from each of the out-trigger boom at depths ranging from 50 to 150 meter. Multi-day fishing trips (20 to 40 days) were undertaken by these trawlers.

Based on the target groups, three types of trawl nets were in operation along the coast of Andhra Pradesh, namely, fin fish trawls targeted for fish, shrimp trawls for shell fish and cephalopod trawls for cuttle fish and squids.

Based on the number of seams, three types of nets were identified, namely, two seam, four seam and six seam. Finfish and cephalopod trawls were two seam trawl; while shrimp trawls had four and six seams.

Generally the trawls with head rope length ranging from 20 to 81 meter were fabricated with polyethylene monofilament twisted twine and were rigged with 50 to 55 kg flat rectangular otter boards

Fish trawls

All fish trawls were made of two seams. Fish trawls of four different sizes, namely, 28, 44 and 46.2 meter targeting different fin fish groups and 81 meter targeting exclusively pomfret were most common types found.

28 meter two seam fish trawl

Design details of this net are given in Fig. 146. The wing, square and first belly of the net was fabricated with 1.25 mm dia high density polyethylene (HDPE) twine of 200 mm mesh size. The upper and lower wing panel and square have depths of 62.5, 80.5 and 18 meshes respectively. The head rope and foot rope were made of 14 mm dia. polypropylene (PP) rope. Seventeen floats of 15 and 25 mm diameter were mounted on head rope. Eighty rubber bobbins weighing 350 gram each were mounted on the foot rope. The net was operated from boats of 13.1 meter LOA fitted with 102 HP engine at a depth range of 20 to 90 meter targeting mainly ribbon fish, upenoides, scianids, lizard fishes and nemipterids.

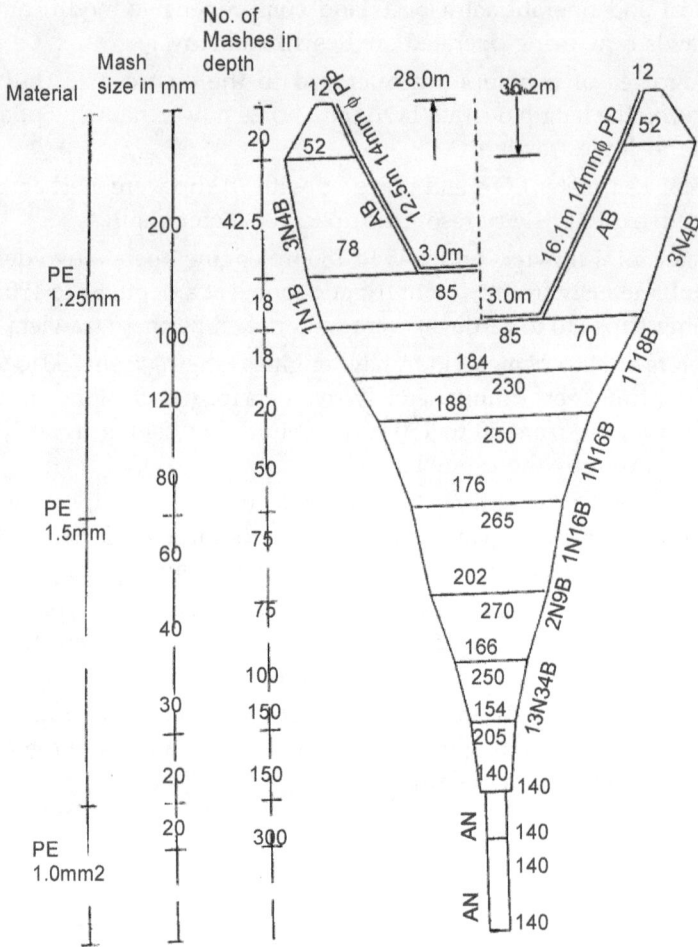

Fig. 146: Design of a typical 28 meter two seam fish trawl.

44 meter two seam fish trawl

The design details of the net are given in Fig. 147. This net was operated from 13.1 meter LOA boat fitted with 102 HP engine at depths ranging from 20 to 90 meter. The wing to first belly of the net was fabricated with 1.5 mm dia, HDPE webbing having 400 mm mesh size. The second and fourth bellies were fabricated with 1.5 mm dia. twine and 5th to 11th bellies with 1.25 mm dia. twine. The upper and lower wing panels had a depth of 51 and 61 meshes. The depth and width of square was 10 x 175 meshes and thai of first belly was 20 x 175 meshes. The cod end was fabricated with 1.5 mm dia. twine having a mesh size of 20 mm. Foot rope and head rope were fabricated with 14 mm dia. polypropylene rope. Seven floats of 51 mm dia. were mounted on head rope and 12 mm iron chain of 45 kg are rigged on foot rope. The leg of the net was 5 meter and the sweep line length was 20 meter.

The net targeted mainly fin fishes like, ribbon fish, upenoides, sciaenids, lizard fishes, nemipterids and silver bellies.

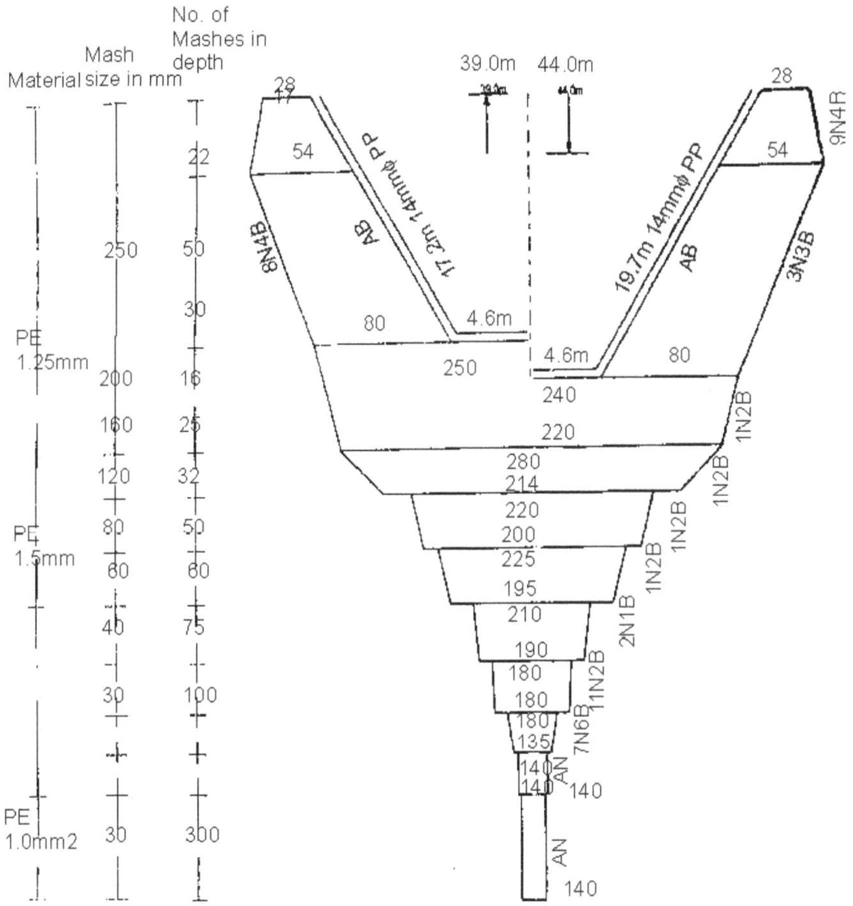

Fig. 147: Design of a typical 44 meter two seam fish trawl

46.2 meter two seam fish trawl

The net operated from 13.1 meter LOA fitted with 102 HP engine at depth ranging from 20 to 90 meter targeted mainly ribbon fish, upenoides, sciaenids, lizard fishes, nemipterids and silver bellies. The wing, square and first belly of the net were fabricated with 400 mm mesh of 1.25 mm dia. twisted HDPE twine (Fig. 148). The upper and lower wing panels had a depth of 64 and 69 meshes respectively. The square had a depth and breadth of 5 x 180 meshes and the first belly was of 9 x 180 meshes. Cod end had a mesh size of 20 mm with depth and breadth of 170 x 140 meshes. The cod end was covered with a rubline having 150 mm mesh size fabricated with 4 mm rope. Seven 400 mm dia. round plastic floats were rigged to head rope of 14 mm dia. and 12 mm iron chain of 40 kg was rigged to foot rope of 14 mm diameter.

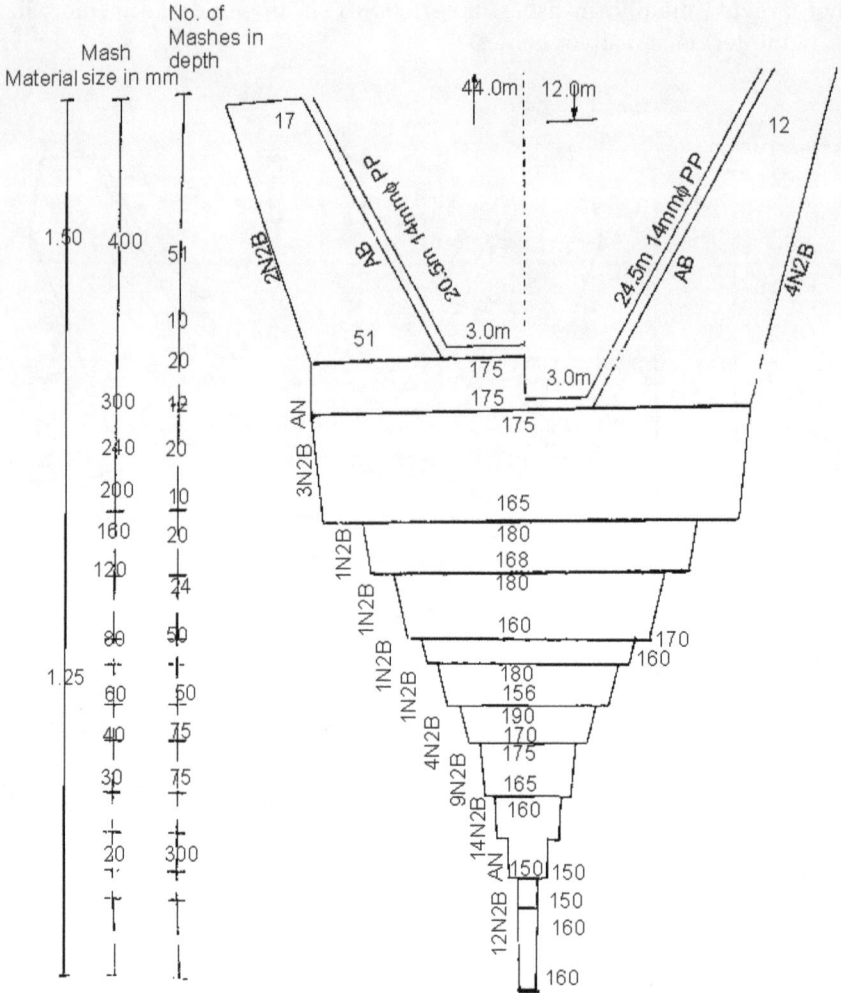

Fig. 148: Design of a typical 46.2 meter two seam fish trawl

81 meter two seam pomfret fish trawl

This net is operated from 13.1 meter LOA fitted with 102 HP engine at depth range of 20 to 90 meter, mainly targeted for pomfret, *(Pampas argenteus, Parastromateus niger)* and other fishes like, ribbon fish, upenoides and lizard fish. The wing, square and first belly of the net were fabricated with 2 mm dia. HDPE webbing of 2000 mm mesh size. The upper and lower wing panels had a depth of 19 and 21 meshes. The depth and width of square was 2x114 meshes, while that of first belly was 4x110 meshes and that of second belly was 4 x 148 meshes and the mesh size was 1500 mm. The third and fourth bellies were fabricated with 1.5 mm diameter twine and the 5th to 14th bellies with 1 mm diameter twine. Cod end was fabricated with 1.5 mm diameter twine having a mesh size of 20 to 25 mm. Foot rope and head rope were of 14 mm diameter polypropylene rope. Five to seven floats of 12 inch dia. spherical plastic floats were attached along the head rope and 8 mm iron chain of 45 to 50 kg were rigged on foot rope. The length of the leg was 1.5 meter on each side and 45 meter sweep line was attached to the net. The net was operated along the current (leeward).

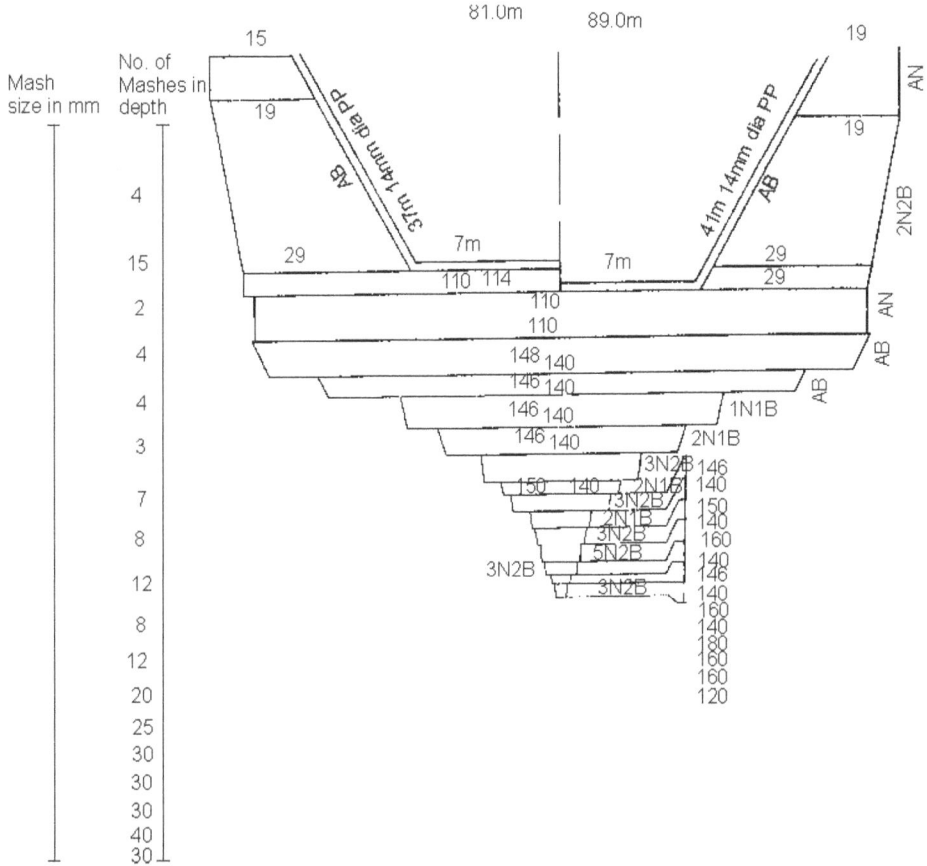

Fig. 149: Design of a typical 81 meter two seam pomfret fish trawl.

Cuttle fish trawl

Only one type of cuttle fish trawl, namely, 39 meter two seam was found in operation. The 39 meter two seam cuttle fish trawl operated by 13.1 meter boat fitted with 102 HP engine at a depth range of 20 to 90 meter, was used to target cuttle fish, chiefly, *Sepia pharaomin* and other fish like, serranids, upenoides and sciaenids. The wing, square and first belly of the net was fabricated with 250 mm mesh size webbing of 1.25 mm diameter twisted HDPE twine. Other parts of the net was fabricated with 1.0 mm twine. Upper and lower wing panels were of 72 and 82 meshes in depth respectively. The square was having a depth of 10 meshes, the breadth of 250 meshes and the mesh size was 250 mm. The cod end was fabricated with 1 mm double twine of 20 to 25 mm mesh size. Eight plastic spherical floats of 400 mm dia. were mounted on the head line. About 84 rubber bobbins each of 500 gram were tied to the ground rope.

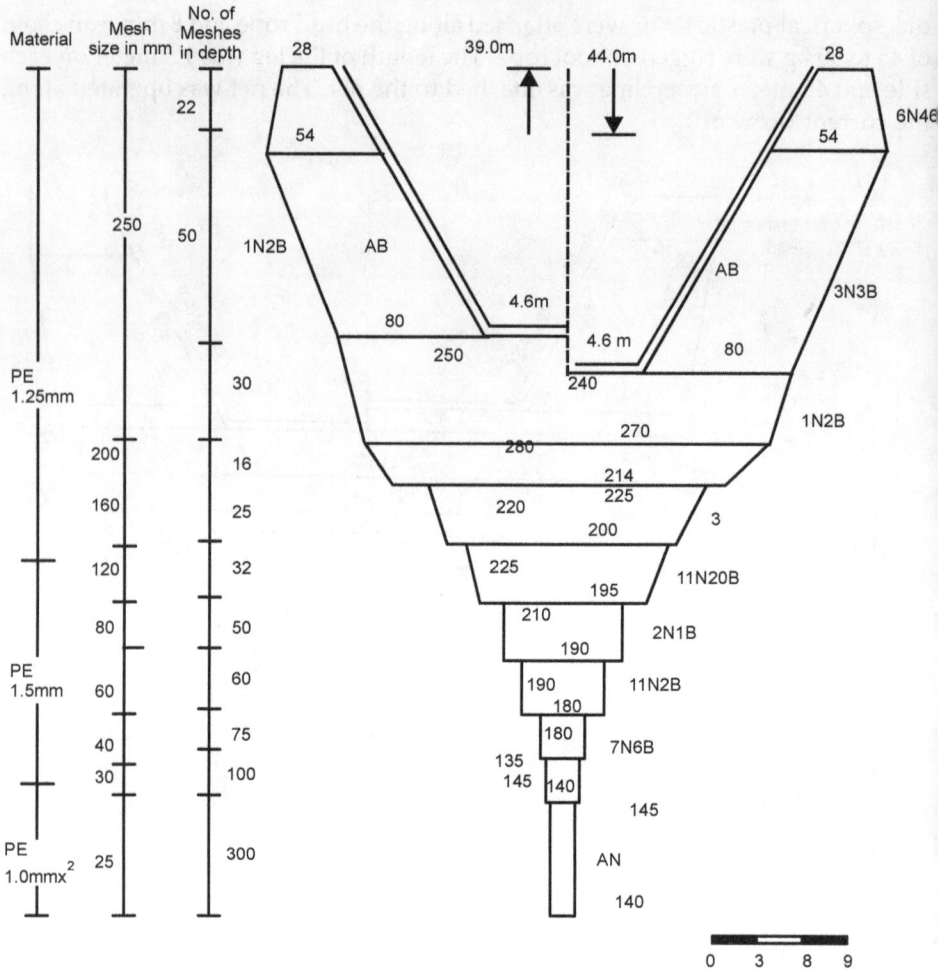

Fig. 150 - Design of a typical 39 meter two seam cuttle fish trawl

27.4 meter two panel six seam shrimp trawl

Design details of a typical 27.4 meter two panel six seam shrimp trawl was depicted in Fig. 151. The six seam shrimp trawl net used from 13.1 meter LOA boat fitted with 102 HP engine was fabricated with two panels and each panel was made with three pieces of webbing. The wing, square and first belly of the net were fabricated with webbing of 0.65 mm twisted polyethylene twine having 60 mm mesh size. The depth, breadth and mesh size of the first belly of the upper panel was 63 x 548 x 60 mm and that of lower panel was 33 x 540 x60 mm (Fig. 151). The net was rigged with flat rectangular wooden otter boards weighing 50 to 55 kg.

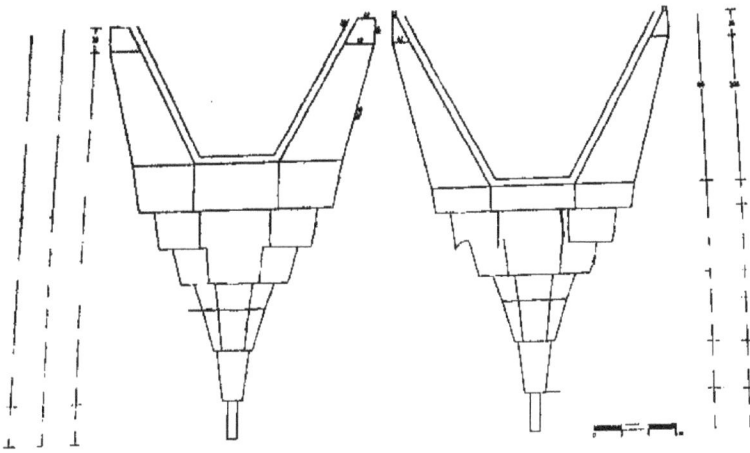

Fig. 151: Design of a typical 27.4 meter two panel six seam shrimp trawl

20 meter four seam shrimp trawl

A four seam 20 meter shrimp trawl was used from mini-trawlers of 16 meter LOA and large trawlers of 21 meter LOA and above. Design details are given in Fig. 152. The net was fabricated with 1.5 mm polyethylene twisted twine. The upper and lower wing panels had 80 meshes depth each. The wing, square and bellies were fabricated with webbing of 50 mm mesh size. Mexican otter boards each weighing 225 kg with sled made of iron of 250 kg were used in the operation. Four nets of 20 meter length each with two nets on each side were operated by a single large steel trawler.

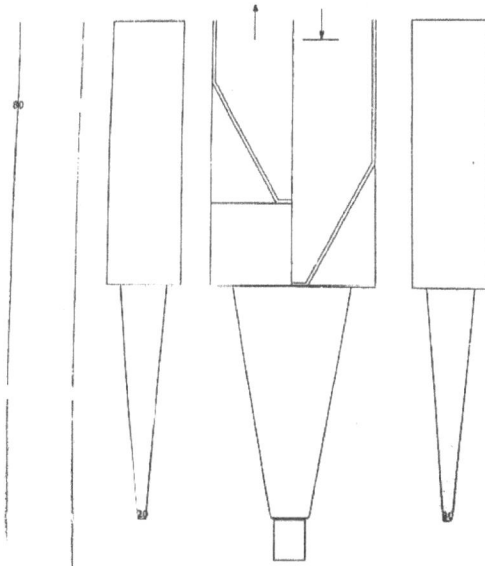

Fig. 152: Design of a typical 20 meter four seam shrimp trawl.

Modification in the trawl sector

There has benn a rapid expansion of both the non-mechanized and mechanized fleet over the past four decades. Narayanappa *at el* (1985) correlated the catch with increase in number of fishing boats and observed reduction trend in the catch per unit of effort (CPUE).

Two seam fish trawls dominated the fishing industry until 1950s, During the next two decades, though popularization of trawling brought about a greater use of four seam nets along the west coast, two seam nets continued its popularity along the east coast. This trend still continues. It was observed that shrimp trawls have undergone several changes in the course of time, like increasing the number of seams from two to six. Results indicated that the mesh size in the fore part of the net increased from 150 mm to 2000 mm in the fish trawls, the advantage being a reduction in the drag.

Four seam and six seam shrimp trawls gained popularity along the east coast. The horizontal opening of conventional two seam trawl was invariably higher than four seam trawl, which indirectly indicated that four seam net obtained more vertical spread, which accounted for better catch of bottom fishes. Further, no significant difference in the catch rate was reported but a striking difference in the catch composition between two seam and four seam trawls was observed. It was evaluated that the double trawl net was effective in targeting bottom as well as off bottom fishes simultaneously. The fuel consumption rate was significantly less in rope trawl while catch was significantly more.

Two seam fish trawls are widely used for exploitation of fin fish resources. Shrimp trawls have undergone several changes in course of time, namely, increase in number of seams from two to six, increase in vertical height and length of net. In order to manage the trawl resources efficiently, caution needs to be exercised in the use of trawl nets. Selective fishing practices and mesh size regulation need to be strictly adhered to, for longer sustainability of fishery resources.

References

Boopendranath, M.K, *et.al...*   Trawl cod end selectivity in respect of silver pomfret *Pampas argenteus* (Euphrasen, 1788); Fishery Technology, 49, 2012

Balasubramanyan, R   A note on the method of reconditioning un serviceable steel otter doors with glass fiber scathing, Fishery Technology, Vol. VIII, No. 2,1971.

Boopendranath, M.R. et *al*   Soft By catch Reduction Device for bottom trawls, a review, Fishery Technology, Vol. 47(2), 2010.

Deshpande, S..D *et al*   Results of preliminary fishing trials with 15.8 m six-seam otter trawl, Fishery Technology, Vol. VII, No. 2,1970

Edwin Leela, *et al*	Trawl cod end selectivity of Torpedo Scad, *Megalapsis cordyla* (Linnaeus, 1758), Fishery Technology, 50 (2013).
Gibinkumar, T.R. *et al*	By catch characterization of shrimp trawl landings off South West Coast of India, Fishery Technology, 49,2012
Juva Charles. J, *et al*	Innovation decision efficiency on selected fishing technologies among steel fishing trawler operators, Fishery Technology, Vol. 48(1),2011.
Kungipalu K.K, *el. al*	Development of trawls for medium sized trawlers for Verabal, North-West coast of India, Fishery Technology, Vol. 16, 1979
Kartha, K.N.	Evolution of suitable of bottom trawls for the medium size steel Trawlers of Orissa Fisheries Department, Fishery Technology Vol. Xin, No. 2, 1976.
Mahalathkar, H.N, G. Jagadeesan	Belly-depth studies for shrimp trawls-part II, Fishery Technology Vol. VII, No. 2, 1970.
Mahathkar, H.N. & G. Jagadeesan	Belly-depth studies for shrimp trawls, Part-I, Fishery Technology Vol. VIE, No. 2. 1971.
Mahalathkar, H.N. & G. Jagadeesan	Belly-depth studies for shrimp trawls, Part-II Fishery Technology Vol. VII,No. 2,1971.
Madhu$_4$ V.R, S.K.Panda B. Meenakumari	Trawl selectivity *onJohniusdussumieri*(Cuvier, 1830) along Gujrat, North West Coast, India, Fishery Technology, 50,2013
Meyer-Waarden, P.F.	Electrical Fishing, FAO Fisheries Study No.7, FAO of United Nations, Rome, 1957.
Mukherjee, D.B	Hydraulics for small fishing trawlers, Fishery Technology, Vol. XIII, No. 2, 1976.
Muthukrishnan, B	Over powering; avoid it; Sea Food Export Journal, Vol. VII, No. 4, 1975.
Main, J & G.I. Sangster	Fish reaction to trawl gear- A study comparing light and heavy ground gear, Report no. 27, Scottish Fisheries Research, Deptt of Agriculture Fisheries for Scotland, 1983.
Neethiselvan, N & G. Brucelee	Analysis of design features offish trawls and shrimp trawls of Thoothukkudi Coast, Fishery Technology, Vol. 40(1), 2003.
Nayar, Gopalan S & R.S. Nair	Recent trends in the design aspects of four seam trawls operated along the South West Coast and South East Coast of India, Proc. on Mechanized Fisheries, 1970.
Nair, R.S. *et al*	Studies on the length of overhang for trawls, Fishery Technology, Vol. VI, No. 1, 1971.

Pillai, N.S, *et al*	Evolution of suitable trawl nets for medium size trawlers, comparative efficiency between 38 m bulged belly, long wing and four panel trawls Fish Technol, Vol. 15,1978.
Prasert Masthawee	Trawl net design, SEAFDC, 1982.
Pillai, N.S. *et al*	Evolution of suitable net for medium sized vessels- Introduction of large meshed high opening fish trawl, Fishery Technology, Vol. 16, No. 2, 1979.
Pravin, P *et al*	Large mesh demersal trawls in India : An update, Fishing Chimes, Vol. 32, No. 5,2012
Pravin, P *et al*	Hard by catch reduction devices for bottom trawls- a review, Fishery Technology, Vol. 48(2), 2011.
Rajeswari, G *et al*	Trawl design used in small scale mechanized fisheries sector of Andhra Pradesh, India, Fishery Technology, 49, 2012.
Raghu Prakash, R *el al*	Size selectivity of square mesh cod ends for *Saurida tumbil* (Bloch 1975) and *Nibea macu late* (Bloch & Schneider, 1801) in Bay of Ben gal, Fishery Technology, 50 (2013)
Swaminath, M *et al*	Trawl gears operated by EFP, Special Publica tion, No.l Exploratory Fisheries Project, Govt. of India, 1979.
SEAFDC	Method of computation of suitable size of trawl gear to horse power of main engine of trawler, 1982.
Sivadas, T.K	Instruments and data acquisition systems for research on trawl nets, Fishery Technology, Vol. VIII, No.2,1971.
Satyanarayana, A.V.V *et al*	Preliminary observations on the design and operation of a three panel double trawl net, Fish Technol, Vol. 13, No. 1, 1976.
Sivan, T.M. *et al*	Some observation on the performance of 10.5 m mid-water trawl operated off Verabal, Fish ery Technology, Vol. VII, No.2,1970.
Sharief, A.T	A brief note on the exploratory bottom trawl operation off Madras, by M.V. Meenabharathi
Satyanarayana, A.V.V *et al*	Further investigations on the relative effi ciency of different shaped otter boards, Fish Technol, Vol. 15,No.l, 1978.
Sivadas, T.K	Instrumentation in fishing gear research, Fish ery Technology, Vol.V, NO. 2, 1970.

Sivadas, T.K	Portable electronic warp load meter, Fish Technol, Vol. 15, No.l,1978
Sabu, S *et al*	Performance of sieve net, By catch Reduction Device in the Seas off Cochin (south West Coast), India, Fishery Technology, 50, 2013.
Verghese, C.P	One boat mid-water trawl-A construction and rigging profile, Bulletin No. 2, Integrated Fisheries Project, 1971.
Verghese, C.P	A note on one-boat mid-water trawling experi ment conducted on the SSW Coast of India, Sea Food Export Jpurnal, Vol. VTI, No.4,1975
Verghese, C.P & P.R. Nair	Development of two boat mid-water trawl ing along Kerala Coast, Sea Food Export Jour nal, Vol. VII, No.9,1975.
Verghese, C.P	One boat mid-water trawl-A construction and rigging profile, Bull. No. 2, Integrated Fish eries Project, 1977.
Verghese, C.P	One boat mid-water trawling experiments conducted on the SSW Coast of India, Sea Food Export Journal, Vol. VII, No.4, 1975.
Vijayan, V *et al*	Target specific 50 m long wing semi-pelagic trawl for off-bottom fishing in Indian EEZ, Fishert Technology, Vol. 40 (1), 2003.
Vijayan, V *et al*	Operational efficiency of Suberkrub and Poly valent otter boards for target specific inshore semi-pelagic trawling, Fishery Technology, Vol 40(10, 2003.
WFA	Fisheries Training Center and Flume Tank, White Fish Authority Development Unit, 1976.
Yamah	Small scale otter trawl, Fishery Journal, No. 17,1982.

Index

A

Apron – 3
Accessories – 47
Authement Ledet Excluder – 178
Andrews soft TED – 179
Angle of attack – 196,197

B

Bosom – 2
Belly – 3, 61, 62, 124
Bolch line – 3
Bating – 4, 7
Bridles – 48
Bobbins – 49
Babylon – 97, 98, 99
Bottom trawl – 111, 114, 116, 117, 118, 120
Bulged belly trawl – 115, 126
B.R.D. 161 – 155, 170, 173
By – catch – 156, 173 – 157
Bent grids BRD – 171, 172
Big-eye BRD – 178

C

Cod end – 3, 63, 143
Creasing – 4, 7
Cutting pattern – 5, 7, 56
Constructionof trawl – 11, 54
Classification – 51
Combination BRD – 174

D

Design – 52, 55, 57, 85, 86
Double trawl – 93
Diamond BRD – 177
Depth recorder – 193

E

Evolution of trawl – 69
Engine HP – 67, 69, 72, 77
Electrical trawl – 185, 187
Electrodes – 180 – 190
Engineering performance – 212

F

Flapper – 3, 64
Foot rope – 11
Fish trawl – 11, 12, 13, 14, 15, 16, 17,27 225-229
Fishing dept – 100
Four panel trawl – 127
Four seam trawl – 129
Funnel BRD – 177
Fish behaviour – 207, 211
Flume tank – 213, 217

G

Granton trawl – 73
Gulf of Mannar – 86-89
Gallows – 228

H

Handing – 9, 53
Head rope relation – 130 – 135
Hydraulic system – 183, 184
Horizontal opening – 199

I

Instrumentation - 83, 84, 192

J

Jibs – 2

L

Lines – 3

Lacing – 8
Long win trawl – 135

M

Mounting – 9
Mid-water trawl – 97, 98, 101, 105, 106, 107, 109
Monofilament BRD – 177
Morrison soft TED – 178
Modification in trawl – 232

N

Neil-Olsen BRD – 177
Nozzles – 218

O

Otter board – 35, 37, 44, 77, 78, 103, 107, 128, 1635, 145, 152, 219
Otter board rigging – 45, 46
Operation of trawl – 218
Otter trawls – 221, 222, 223, 234

P

Parts of trawl – 2, 51
Panels – 2
Pnnants – 48
Preservation – 49
Performance – 82
Pelagic trawl – 108
Parker soft TED – 179
Power requirement – 181

R

Rigging – 84, 104, 108, 120, 128
Rope BRD – 176

S

Square – 4, 58, 95, 96
Sewing – 8
Shrimp trawl – 28, 34, 36, 91, 124, 230, 231

Sinker – 48
Shape – 68
Six seam – 74, 92
Semi-pelagic trawl – 121, 122
Silver pomfret – 140
Square mesh – 141
Sieve trawl – 153, 154, 179
Soft by-catchj devices – 175
Square mesh window – 175

T

Throat – 9
Two seam trawl – 13
Take up – 9
Trawl gear – 10, 14
Tickler chain – 49
Two seam – 74
Transducer – 102
Technology transfer – 136
Trawl selectivity – 13
Trend – 178
Telmetering instruments – 201
Terminology for trawl gear – 202, 206

V

V-cut BRD – 178
Vertical opening – 198

W

Webbings – 2, 49
Wings – 2, 59, 60
Warp – 48, 218
Warp tension – 194, 195
Water flow – 200
Water current – 200
Winch – 219